Applications of Field-Programmable Gate Arrays in Scientific Research

Applications of Field-Programmable Gate Arrays in Scientific Research

Hartmut F.-W. Sadrozinski
University of California
Santa Cruz, USA

Jinyuan Wu
Fermi National Accelerator Laboratory
Batavia, Illinois, USA

CRC Press
Taylor & Francis Group
Boca Raton London New York

CRC Press is an imprint of the
Taylor & Francis Group, an **informa** business

CRC Press
Taylor & Francis Group
6000 Broken Sound Parkway NW, Suite 300
Boca Raton, FL 33487-2742

First issued in paperback 2017

© 2011 by Taylor and Francis Group, LLC
CRC Press is an imprint of Taylor & Francis Group, an Informa business

No claim to original U.S. Government works

ISBN-13: 978-1-4398-4133-4 (hbk)
ISBN-13: 978-1-138-11248-3 (pbk)

Visit the Taylor & Francis Web site at
http://www.taylorandfrancis.com

and the CRC Press Web site at
http://www.crcpress.com

Contents

Preface...ix
Acknowledgments xi
The authors xiii

Chapter 1   Introduction 1
1.1   What is an FPGA?...1
1.2   Digital and analog signal processing 1
1.3   FPGA costs...1
1.4   FPGA versus ASIC..3
References 4

Chapter 2   Understanding FPGA resources 5
2.1   General-purpose resources 5
    2.1.1   Logic elements...5
    2.1.2   RAM blocks 6
2.2   Special-purpose resources..7
    2.2.1   Multipliers..7
    2.2.2   Microprocessors..7
    2.2.3   High-speed serial transceivers 8
2.3   The company- or family-specific resources 8
    2.3.1   Distributed RAM and shift registers 8
    2.3.2   MUX..8
    2.3.3   Content-addressable memory (CAM)...9
References 9

**Chapter 3   Several principles and methods of resource usage
             control**..11
3.1   Reusing silicon resources by process sequencing 11
3.2   Finding algorithms with less computation.......................................12
3.3   Using dedicated resources 13
3.4   Minimizing supporting resources..14
    3.4.1   An example...14
    3.4.2   Remarks on tri-state buses 14

3.5   Remaining in control of the compilers 16
    3.5.1   Monitoring compiler reports on resource usage and
            operating frequency 16
    3.5.2   Preventing useful logic from being synthesized away
            by the compiler 16
    3.5.3   Applying location constraints to help improve
            operating frequency 18
3.6   Guideline on pipeline staging 18
3.7   Using good libraries 19
References 20

Chapter 4   Examples of an FPGA in daily design jobs 21
4.1   LED illumination 21
    4.1.1   LED rhythm control 21
    4.1.2   Variation of LED brightness 23
    4.1.3   Exponential drop of LED brightness 23
4.2   Simple sequence control with counters 24
    4.2.1   Single-layer loops 25
    4.2.2   Multilayer loops 27
4.3   Histogram booking 31
    4.3.1   Essential operations of histogram booking 31
    4.3.2   Histograms with fast booking capability 33
    4.3.3   Histograms with fast resetting capability 35
4.4   Temperature digitization of TMP03/04 devices 37
4.5   Silicon serial number (DS2401) readout 38
References 41

Chapter 5   The ADC + FPGA structure 43
5.1   Preparing signals for the ADC 43
    5.1.1   Antialiasing low-pass filtering 43
    5.1.2   Dithering 44
5.2   Topics on averages 46
    5.2.1   From sum to average 46
    5.2.2   Gain on measurement precision 46
    5.2.3   Weighted average 47
    5.2.4   Exponentially weighted average 48
5.3   Simple digital filters 50
    5.3.1   Sliding sum and sliding average 51
    5.3.2   The CIC-1 and CIC-2 filters 52
5.4   Simple data compression schemes 53
    5.4.1   Decimation and the decimation filters 53
    5.4.2   The Huffman coding scheme 55
    5.4.3   Noise sensitivity of Huffman coding 56
References 57

Chapter 6   Examples of FPGA in front-end electronics......................... 59
6.1   TDC in an FPGA based on multiple-phase clocks........................... 59
6.2   TDC in an FPGA based on delay chains.. 62
        6.2.1   Delay chains in an FPGA 63
        6.2.2   Automatic calibration... 64
        6.2.3   The wave union TDC 67
6.3   Common timing reference distribution 69
        6.3.1   Common start/stop signals and common burst 69
        6.3.2   The mean timing scheme of common time reference........ 70
6.4   ADC implemented with an FPGA 70
        6.4.1   The single slope ADC.. 71
        6.4.2   The sigma-delta ADC... 73
6.5   DAC implemented with an FPGA 74
        6.5.1   Pulse width approach 74
        6.5.2   Pulse density approach... 75
6.6   Zero-suppression and time stamp assignment................................ 77
6.7   Pipeline versus FIFO 78
6.8   Clock-command combined carrier coding (C5)............................... 82
        6.8.1   The C5 pulses and pulse trains 82
        6.8.2   The decoder of C5 implemented in an FPGA.................. 83
        6.8.3   Supporting front-end circuit via differential pairs............ 85
6.9   Parasitic event building 86
6.10   Digital phase follower 88
6.11   Multichannel deserialization 92
References 95

Chapter 7   Examples of an FPGA in advanced trigger systems 97
7.1   Trigger primitive creation 97
7.2   Unrolling nested-loops, doublet finding.. 99
        7.2.1   Functional block arrays.. 100
        7.2.2   Content-addressable memory (CAM)....................... 102
        7.2.3   Hash sorter 105
7.3   Unrolling nested loops, triplet finding.. 106
        7.3.1   The Hough transform 108
        7.3.2   The tiny triplet finder (TTF).................................... 110
7.4   Track fitter.. 110
References 114

Chapter 8   Examples of an FPGA computation 115
8.1   Pedestal and RMS.. 115
8.2   Center of gravity method of pulse time calculation....................... 116
8.3   Lookup table usage.. 118
        8.3.1   Resource awareness in lookup table implementation....... 118
        8.3.2   An application example 119

8.4   The enclosed loop microsequencer (ELMS).................................. 122
References 124

Chapter 9   Radiation issues 125
9.1   Radiation effects 125
       9.1.1   TID 125
       9.1.2   SEE effects... 125
9.2   FPGA applications with radiation issues 126
       9.2.1   Accelerator-based science.. 126
       9.2.2   Space 126
9.3   SEE rates.. 127
9.4   Special advantages and vulnerability of FPGAs in space 128
9.5   Mitigation of SEU... 129
       9.5.1   Triple modular redundant (TMR) 129
       9.5.2   Scrubbing.. 129
       9.5.3   Software mitigation: EDAC... 129
       9.5.4   Partial reconfiguration.. 130
References 130

**Chapter 10 Time-over-threshold: The embedded particle-
                tracking silicon microscope (EPTSM) 131**
10.1   EPTSM system... 131
10.2   Time-over-threshold (TOT): analog ASIC PMFE 133
10.3   Parallel-to-serial conversion... 135
10.4   FPGA function 135
References 137
Appendix: Acronyms 139
Index 143

Preface

Outline of the book

The book is an introduction to applications of field-programmable gate arrays (FPGAs) in various fields of research. It covers the principle of the FPGAs and their functionality. The main thrust is to give examples of applications, which range from small one-chip laboratory systems to large-scale applications in "big science." They give testimony to the popularity of the FPGA system.

A primary topic of this book is resource awareness in FPGA design. The materials are organized into several chapters:

- Understanding FPGA resources (Chapter 2)
- Several principles and methods (Chapter 3)
- Examples from applications in high-energy physics (HEP), space, and radiobiology (Chapters 4–10)

There is no attempt made to identify "golden" design rules that will be sure choices for saving silicon resources. Instead, the purpose of this book is to remind the designers to pay attention to resources at the planning, design, and implementation stages of an FPGA application. Based on long experience, resource awareness considerations may slightly add to the load of designers' brain work and sometimes may slightly slow down the development pace, but its saving in silicon resources and therefore direct and indirect cost is significant.

Philosophy of this book

This book contains many hands-on examples taken from many different fields the authors have been working in. Its emphasis is less on the computer engineering details than on concepts and practical "how-to." Based on the (sometimes painful!) experiences of the authors, sound design practices will be emphasized. The reader will be reminded constantly during the discussion of the sample applications that the resources of the FPGA are limited and need to be used prudently. The authors want to influence the design habit of the younger readers so that they keep in mind savings of silicon resources and power consumption during their design practice.

Target audience

The book targets advanced students and researchers, who are interested in using FPGAs in small-scale laboratory applications, replacing commercial data acquisition systems with fixed protocols with flexible and low-cost alternatives. They will find a quick overview as to what is possible when FPGAs are used in data acquisition, signal processing, and transmission. In addition, the general public with an interest in the potential of available technologies, will get a very wide-angle snapshot of what that "buzz" is all about.

Use of this book

The book may serve as a supplementary reading in digital design classes (CE, EE) and instrumentation classes (physics). Some examples presented in the book can be used for student laboratories.

Additional supporting material

There is always the question of how much practical knowledge can be transferred in a printed book. In order to supply much more detail of FPGA programming and usage, the authors are maintaining a Web site (http://scipp.ucsc.edu/~hartmut/FPGA) containing design details of the study cases mentioned in the book—for example selected VHDL code, detailed schematics of selected projects, photographs and screen shots, etc., that are not suitable for a hard-copy book.

Acknowledgments

HFWS would like to thank his colleagues Ned Spencer, Brian Keeney, Kunal Arya, Ford Hurley, Brian Colby, and Eric Susskind for their valuable contributions and comments. JYW would wish to thank his colleagues and friends Robert DeMaat, Sten Hansen, Tiehui Liu of Fermilab, Fukun Tang of the University of Chicago, William Moses, Seng Choong of the Lawrence Berkeley Lab, and Yun Wu of Apple, Inc. for their valuable contributions over the years.

The authors

Hartmut F.-W. Sadrozinski has been working on the application of silicon sensors and front-end electronics for the last 30 years in elementary particle physics and astrophysics. In addition to getting very large detector systems planned, built, tested, and operated, he is working on the application of these sensors in the support of hadron therapy.

Jinyuan Wu received his BS degree in Space Physics from Department of Geophysics, Peking University, Beijing, China, in 1982; MS degree in Micro-ElectroUltrasonic Devices from Institute of Acoustics, Chinese Academy of Sciences, Beijing, China, in 1986; and Ph.D. degree in Experimental High Energy Physics from the Department of Physics, The Pennsylvania State University in 1992. He has been an Electronics Engineer II and III in the Particle Physics Division, Fermi National Accelerator Laboratory since 1997. He is a frequent lecturer at international workshops and in IEEE conferences refresher courses.

chapter one

Introduction

1.1   What is an FPGA?

An FPGA (field-programmable gate array) consists of logic blocks of digital circuitry that can be configured "in the field" by the user to perform the desired functions. In addition, it contains a set of diverse service blocks such as memories and input/output drivers. In contrast to application-specific integrated circuits (ASICs; "chips"), which are designed to fulfill specific predetermined functions, FPGAs are "fit-all" devices that provide a generalized hardware platform that can be configured (and reconfigured unlimited times over) by downloading the firmware tailored to the function. The power of FPGAs can be traced to their ability to perform parallel functions simultaneously, and to the fact that they contain digital clock management functions supplying several high-speed clocks. With the advantage of much larger freedom and wealth of opportunity comes the disadvantage of limited and predetermined resources (RAM, etc).

1.2   Digital and analog signal processing

FPGA applications have been very popular in high-energy/nuclear physics experiment instrumentation. The functionalities of the FPGA devices range from merely glue-logic to full data acquisition and processing. One reason for the popularity of FPGAs is that although they are by nature digital devices, they can be used to process analog signals if the signal can be correlated with time. Given that FPGAs support low-noise data transmission through low-voltage digital signal (LVDS) protocols, they are in the center of many mixed-signal applications. They allow moving analog information quickly into the digital realm, where the signal processing is efficient, fast, and flexible. Examples are pulse height analysis through charge-to-time converters and time-over-threshold counters.

1.3   FPGA costs

Similar to any computing options, the FPGA computing consumes resources. The direct resource consumption is essentially in terms of silicon area, which translates into the cost of the FPGA devices. As a result of direct silicon resource consumption, indirect cost must also be paid in terms of FPGA recompile time, printed circuit board complexity, power usage, cooling issues, etc.

There is folklore that "FPGAs are cheap." This is certainly true when comparing the cost of small numbers of FPGAs with small numbers of ASICs. The actual prices of several Altera FPGA device families taken from the Web site of an electronic parts distributor (Digi-Key Co. [1], May 25, 2010) are plotted in Figure 1.1. Each device may have various speed grades and packages, the prices of which vary greatly. The lowest price for each device is chosen for our plots.

It can be seen that FPGA devices are not necessarily cheap. In terms of absolute cost, there are devices costing as little as $12, and there are also devices costing more than $10,000. When compared within a family, lower–middle sized devices have the lowest price per logic element, as shown in Figure 1.2.

Another fact that must be mentioned here is that the FPGA design is not a "program," even though the design can be in the format of "code" of languages such as VHDL. The FPGA design is a description of a circuit that is configured and interconnected to perform certain functions. A line of the code usually occupies some logic elements, no matter how rarely it is used. This is in contrast to computer software programs, which do not take execution time unless used. In addition, storage of even very large programs in computer memory is relatively cheap in

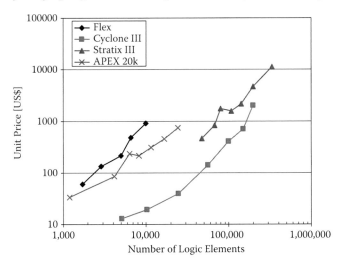

Figure 1.1 Unit price of several Altera FPGA device families as a function of the number of logic elements (extracted from the Digi-Key Co., catalog Web site http://www.digikey.com, May 2010).

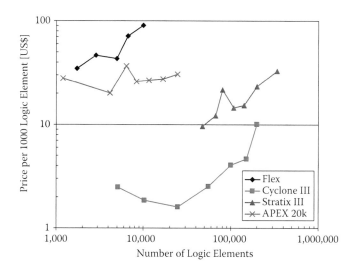

Figure 1.2 Price per 1000 logic elements of several Altera FPGA device families as a function of the number of logic elements (extracted from the Digi-Key Co., catalog Web site http://www.digikey.com, May 2010).

terms of system resources. Therefore, it is a good practice to think and rethink the efficiency of each line in the code during the design. Rarely used functions should be reorganized so that they are performed in the resources shared with other functions as much as possible.

Code reuse is an important trend in FPGA computing, just as in its counterpart of microprocessor computing. Designers should keep in mind that a functional block designed today might be reused thousands of times in the future. Today's design could become our library or intellectual property. If a block is designed slightly too big than needed, it will be too big in thousands of applications in future projects.

What is even worse is that we may learn the wrong lessons from these poor designs. The fear that the firmware will not fit causes planners to reserve excessive costly FPGA resources on printed circuit boards. It is also possible that functions can be mistakenly considered too hard to be implemented in FPGA, resulting in decisions to either degrade system performance or to increase the complexity of system architecture.

1.4   FPGA versus ASIC

The FPGA cost can be studied by comparing the number of transistors needed to implement certain functions in FPGA and non-FPGA IC chips, such as in microprocessors. Several commonly used digital processing functions are compared in Table 1.1.

Table 1.1 Number of Transistors Needed for Various Functions

	Number of transistors	Notes
4-in NAND gate	8	Same for 4-in NOR
Full Adder	24-28	
Static RAM bit	6	Bit storage cell only
FPGA Lookup Table (LUT)	>96	16 storage cells only

Combinational logic functions are implemented with 4-input LUT in the FPGA. The contents of an LUT may be programmed so that it represents a function as simple as a 4-input NAND/NOR or as complicated as a full adder bit with carry supports. In both cases, FPGA uses far more transistors than non-FPGA IC chips. This is the cost one needs to pay for the flexibility one has in configuring an FPGA. Due to this flexibility, FPGA designers enjoy fast turnaround time of design revisions, and lower cost—compared with ASIC approaches—when the number of chips in the final system is small. On the other hand, this comparison tells us that eliminating unnecessary functions in an FPGA saves more transistors than in non-FPGA chips such as ASICs.

References

1. Digi-Key Corporation, catalog Web site http://www.digikey.com, May 2010.

chapter two

Understanding FPGA resources

In this chapter, we use the Altera Cyclone II [1] and Xilinx Spartan-6 families [2] as our primary examples. We break the FPGA resources into several categories, that is, general-purpose resources such as logic elements and RAM blocks, special-purpose ones such as multipliers, high-speed serial communication and microprocessors, and family- or company-specific resources such as distributed RAM, MUX, CAM, etc.

2.1   General-purpose resources

Nearly all RAM-based FPGA devices contain logic elements (logic cells) and memory blocks. These are primary building blocks for the vast majority of logic functions.

2.1.1   Logic elements

The logic elements (LEs) are the essential building blocks in FPGA devices. A logic element normally consists of a 4-input (up to 6 inputs in some families) lookup table (LUT) for combinational logic and a flip-flop (FF) for sequential operation. Typical configurations of logic elements are shown in Figure 2.1.

Usually, logic elements are organized in arrays, and chained interconnections are provided. Perhaps the most common chain support is the carry chain, which allows the LE to be used as a bit in an adder or a counter.

The LUT itself is a small 16×1-bit RAM with contents preloaded at the configuration stage. Clearly, any combinational logic with four input signals can be implemented, which is the primary reason for the flexibility of the FPGA devices. But when more than four signals participate in the logic function, more layers of LUT are normally necessary. For example, if we need a 7-input AND gate, it can be implemented with two cascaded lookup tables.

The output of the combinational signals is often registered by the FF to implement sequential functions such as accumulator, counter, or any pipelined processing stage.

The FF in the logic element can be bypassed so that the combinational output is sent out directly to other logic elements to form logic functions that need more than four inputs. In this case, the FF itself can be used as a "packed register," that is, a register without the LUT.

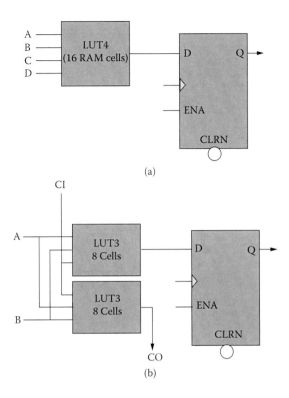

Figure 2.1 Typical configurations of logic elements: (a) normal mode, (b) arithmetic mode.

Just as in any digital circuit design, for a given logic function, the greater the number of pipeline stages, the less the combinational propagation delay between the registers of the stages, and the faster the system clock can operate. Unlike in ASIC, adding pipeline stages in the FPGA normally will not increase logic element usage much, since the FF exists already in each logic element. In practice, however, the number of pipeline stages or the maximum operating frequency is not designed to the maximum value, but rather to a value that balances various considerations.

The logic elements are typically designed to support a carry chain so that a full adder can be implemented with one logic element (otherwise it needs two). Counters and accumulators are implemented with a full adder feeding a register.

2.1.2   *RAM blocks*

RAM blocks are provided in nearly all FPGA devices. In most families, the address, data, and control ports of RAM blocks are registered for synchronous operation so that the RAM blocks can run at a higher speed. It is very common that the RAM blocks provided in FPGA are true dual-port RAM blocks.

If a RAM block is preloaded with initial contents and not overwritten by the users, it becomes a ROM. It is more economical to implement ROM using RAM blocks if a relatively large number of words is to be stored. To implement ROM with fewer than 16 words, use LUT.

The input and output data ports can have different widths. This feature allows the user to buffer parallel data and send out data serially or to store data from a serial port and read out the entire word later.

2.2   Special-purpose resources

In principle, almost all digital logic circuits can be built with logic elements. However, as pointed out earlier, logic elements use more transistors to implement logic functions, which is a trade-off in flexibility. In FPGA devices, certain special-purpose resources are provided so that functions can be implemented with a reasonable amount of resources. For data-flow-intensive applications, specially designed high-speed serial transceivers are provided in some FPGA families for fast communications.

2.2.1   Multipliers

Multipliers become popular in today's FPGA families. Typical multipliers use $O(N^2)$ full adders, where N is the number of bits of the two operands, which would use too many transistors and consume too much power if implemented with logic cells. Therefore, it is recommended to use dedicated multipliers rather than building them from logic cells when the multiplication operations are needed.

However, multiplications are intrinsically resource- and power-consuming operations. If multiplications can be eliminated, reduced, or replaced, it is recommended to do so.

2.2.2   Microprocessors

The PowerPC blocks are found in the Xilinx Virtex-II Pro family [3]. Generally speaking, dedicated microprocessor blocks use fewer transistors compared to implementing the processors with soft cores that use logic elements.

Using microprocessors, either dedicated blocks or soft cores needs to be carefully considered in the planning stage since it is a relatively large investment.

2.2.3   *High-speed serial transceivers*

High-speed serial transceivers are found in both the Altera and Xilinx FPGA families. These transceivers operate at multi-Gb/s data rate, and popular encoding schemes such as 8B/10B and 64B/66B are usually supported. The usefulness of the high-speed serial data links is obvious.

The only reminder for the designers is that the data rate of multi-Gb/s exceeds the needs of many typical data communication links in daily projects. If in a project, a 500 Mb/s or lower data rate is sufficient, it is not recommended to get the "free" "safety factor" to go multi-Gb/s. In addition to the device cost and power consumption, the connectors and cables for multi-Gb/s links require more careful selection and design, while for low-rate links, low-cost twisted pair cables usually work well.

2.3   *The company- or family-specific resources*

Several useful resources can be found in certain FPGA families. These are now described.

2.3.1   *Distributed RAM and shift registers*

The LUTs in the FPGA are typically 16×1-bit RAMs. However, the LUTs are normally written in the FPGA configuration stage, and users cannot modify the contents during the operating stage. In several families of Xilinx FPGA, the LUT can be configured as RAM or shift register so that the user can store information in it. See the application notes [4] and [5] for details.

The distributed RAMs can be used to implement register files. In this case, a logic element stores 16 bits data rather than the 1 bit in typical implementations.

With user-writeable support, the applications of the distributed RAM and shift register are far broader than just storing information. An example of the distributed RAM application can be found in Reference [6]. Another example given in the application note [7] shows the application of the shift register.

2.3.2   *MUX*

In some families of Xilinx FPGA, dedicated multiplexers are designed in addition to the regular combinational LUT logic. A 2:1 multiplexer can certainly be implemented with a regular LUT using three inputs, but a dedicated MUX uses a lot fewer transistors.

When a relatively wide MUX is needed, using dedicated MUX in Xilinx FPGA saves resources when compared with purely using LUT. The application note [8] is a good source of information on this topic.

2.3.3   Content-addressable memory (CAM)

Content-addressable memory is a device that provides an address where the stored content matches the input data. The CAM is useful for the backward searching operation. The Altera APEX II family [9] provides embedded system blocks (ESBs) that can be used as either a dual-port RAM or a CAM. This is a fairly efficient CAM implementation in FPGA devices.

In other FPGA families, normally there is no resource that can be used as a CAM directly. In principle, the CAM function can be implemented with logic elements. However, it is not recommended to build CAM with a wide data port using logic elements since it takes a large amount of resources. Alternatives such as "Hash Sorters" for backward searching functions are more resource friendly.

References

1. Altera Corporation, Cyclone II Device Handbook, 2007, available via: http://www.altera.com/.
2. Xilinx Inc., Spartan-6 Family Overview, 2010, available via: http://www.xilinx.com/.
3. Xilinx Inc., Virtex-II Pro and Virtex-II Pro X Platform FPGAs, 2007, available via: http://www.xilinx.com/.
4. Xilinx Inc., Using Look-Up Tables as Distributed RAM in Spartan-3 Generation FPGAs, 2005, available via: http://www.xilinx.com/.
5. Xilinx Inc., Using Look-Up Tables as Shift Registers (SRL16) in Spartan-3 Generation FPGAs, 2010, available via: http://www.xilinx.com/.
6. J. Wu et al., The application of Tiny Triplet Finder (TTF) in BTeV Pixel Trigger, *IEEE Trans. Nucl. Sci.* vol. 53, no. 3, pp. 671–676, June 2006.
7. Xilinx Inc., Serial-to-Parallel Converter, 2004, available via: http://www.xilinx.com/.
8. Xilinx Inc., Using Dedicated Multiplexers in Spartan-3 Generation FPGAs, 2005, available via: http://www.xilinx.com/.
9. Altera Corporation, APEX II Programmable Logic Device Family, 2002, available via: http://www.altera.com/.

chapter three

Several principles and methods of resource usage control

3.1 Reusing silicon resources by process sequencing

Assuming that there are N computations to be done, each taking one clock cycle in a processing unit (PU), the N computations can be done in $T = N$ clock cycles with one PU. It is also possible to share the computations among M PUs so that the computation can be done in a shorter time; that is, $T = N/M$ clock cycles. If $M = N$ PUs are designed, then all the computation can be done in one clock cycle. This simple resource–time trading-off principle is probably one of the most useful methods of resource usage control in digital circuit design.

A microprocessor contains one processing unit, which is normally called the arithmetic logic unit (ALU). The ALU performs a very simple operation each clock cycle. But within many clock cycles, many operations are performed in the same ALU, and a very complex function can be achieved.

On the other hand, many FPGA functions tend to have a "flat" design, that is, having multiple processing units and performing multiple operations each clock cycle. The flat design allows fast processing but uses more logic elements. If there are several clock cycles in the FPGA between two input data, one may consider using fewer processing units and letting each unit perform several operations sequentially.

Consider the example of finding coincidence between two detector planes, as shown in Figure 3.1. A flat design would need implementing coincidence logic for all detector elements. Assuming that the sampling rate of the detector element is 40 MHz and the FPGA can run at a clock speed of 160 MHz, then only 1/4 of silicon resource for the coincidence unit would be needed. The coincidence between 1/4 of the detector planes can be performed in each clock cycle, and the entire plane can be processed in four clock cycles.

In modern HEP DAQ/trigger systems, digitized detector data are typically sent out from the front-end in serial data links, and the availability of the data is already sequential. It would be both convenient and resource saving to process the data in a sequential manner.

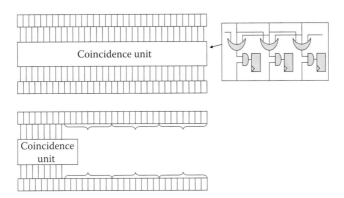

Figure 3.1 An example of process sequencing.

3.2   *Finding algorithms with less computation*

Process sequencing reduces silicon resource usage at the cost of a lower throughput rate when the total number of computations is a constant. When the data throughput rate is known to be low, reducing logic element consumption is a good idea. However, the most fundamental means of resource saving is to reduce the total computations required for a given processing function.

As an example, consider fitting a curved track with hits in several detector planes. To calculate all parameters of a curved track projection, at least 3 points are needed. With more than 3 hit points, the user may take advantage of redundant measurements to perform track fitting to reduce errors of the calculated parameters. The track fitting generally needs additions, subtractions, multiplications, and divisions. However, by carefully choosing coefficients in the fitting matrix, it is possible to eliminate divisions and full multiplications, leaving only additions, subtractions, and bit-shifts. In Ref. 1, this fitting method is discussed.

Another example is the Tiny Triplet Finder (TTF) [2], which groups three or more hits to form a track segment with two free parameters. These processes need three nested loops if implemented in software using $O(n^3)$ executing time, where n is the number of hits in an event. It is possible to build an FPGA track segment finder with execution time reduced to $O(n)$, essentially to find one track segment in each operation. However, typically the track segment finders consume $O(N^2)$ logic elements, where N is the number of bins that the detector plane is divided into. The TTF we developed consumes only $O(N*\log N)$ logic elements, which is significantly smaller than $O(N^2)$ when N is large.

Since the number of clock cycles for execution and silicon resource is more or less interchangeable, fast algorithms developed for sequential computing software usually can be "ported" to the FPGA world, resulting in resource saving. A well-known example is the Fast Fourier Transform (FFT), which exhibits both computing time saving in software and silicon resource saving in the FPGA.

3.3   Using dedicated resources

In the FPGA, a logic element "can do anything." However, more transistors are needed to support the ultraflexibility of the logic elements. In today's FPGA families, resources for dedicated functions with more efficient usage of transistors are provided. Appropriately utilizing these resources helps users to design compact FPGA functions.

Each logic element contains a flip-flop that can be used to store one bit of data. If many words are stored in logic elements, the entire FPGA can be filled up very quickly. Large amounts of not-so-frequently-used data should be stored in RAM blocks. Logic elements should only be used to store frequently accessed data, which are equivalent to registers in microprocessors.

When the data is to be accessed by I/O ports, microcontroller buses, etc., RAM blocks are more suitable. Since the data are distributed to and merged from storage cells inside the RAM blocks, implementing this function outside of the RAM would waste a large amount of resources.

RAM blocks can also be used for purposes other than data storage. For example, very complex multi-inputs/outputs logic functions can be implemented with RAM blocks.

For the fast calculation of square, square root, logarithm, etc., of a variable, it is often convenient to use a RAM block as a lookup table.

The RAM blocks in many FPGA families are dual-port, and the users are allowed to specify different data widths for the two ports. Sometimes, the data width of a port is specified to be 1-bit, which allows the user to make handy serial-to-parallel or parallel-to-serial conversions.

As mentioned earlier, a CMOS full adder uses 24–28 transistors, while an LE in the FPGA takes more than 96 transistors. To implement functions such as counters or accumulators using LE, the inefficiency problem of transistor usage is not very serious. For example, a 32-bit accumulator uses 33–35 logic elements, which are a relatively small fraction of typical FPGA devices. To implement a 32-bit multiplier, on the other hand, at least 512 full adders are needed, and it becomes a concern in applications in which many multiplications are anticipated. Therefore, many today's FPGA families provide multipliers.

Generally speaking, when a multiplication is absolutely needed, it is advisable to use a dedicated multiplier rather than implementing the multiplier with logic elements.

However, there are a finite number of multipliers in a given FPGA device. Multiplication is intrinsically a power-consuming operation, despite relatively efficient transistor usage in dedicated multipliers. Avoiding multiplication or substituting it with other operations such as shifting and addition is still a good practice.

3.4   Minimizing supporting resources

Sometimes, silicon resources are designed not to perform the necessary process, but just to support other functional blocks so that they can process data more "efficiently." In this situation, the supporting logic may require too much of a resource, so that a less efficient implementation might be preferable.

3.4.1   An example

Consider an example as shown in Figure 3.2, in which data from four input channels are to be accumulated to four registers A, B, C, and D, respectively.

The block diagram shown in Figure 3.2a has one adder with data fed by multiplexers that merge 4 channels of sources each. The logic operates sequentially, adding one channel per clock cycle and, clearly, the adder in this case is utilized very "efficiently." However, the 4-to-1 multiplexer typically uses 3 logic elements per bit in the FPGA while the full adder uses only 1 LE per bit. So, an alternative diagram shown in Figure 3.2b actually uses fewer logic elements than the diagram in Figure 3.2a (diagram in Figure 3.2a: 11 LEs/bit, in Figure 3.2b: 4 LEs/bit), although the adders in this case are utilized less efficiently.

Therefore, the principle of process sequencing discussed earlier should not be pushed too far. In actual design, the choice of parallel or sequential design should be balanced with the resource usage of the supporting logic. In the case shown in Figure 3.2c, for example, when the accumulated results are to be stored in a RAM block, sequential design becomes preferable again.

3.4.2   Remarks on tri-state buses

Tri-state buses are common data paths utilized at the board and crate level when multiple data sources and destinations are to be connected together. The buses are shared among all data source devices, and at any given moment only one device is allowed to drive the bus; the output buffers of all other devices are set to a high-impedance (Z) state.

In a broad range of FPGA families, tri-state buffers are only available for external I/O pins but not inside the FPGA. To support porting of legacy system-level designs into an FPGA, the design software provided by vendors usually allows tri-state buses as legal design entry elements. In actual implementation, however, they are converted into multiplexers,

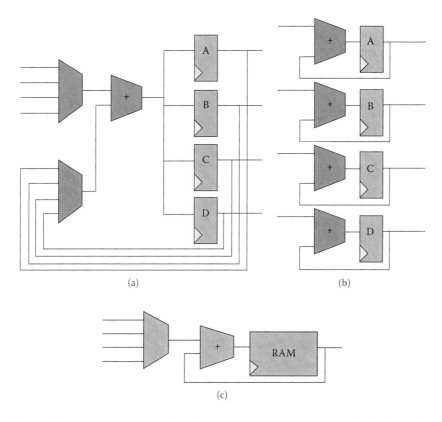

Figure 3.2 An example of minimizing supporting resources: (a) single adder serving four registers, (b) four adders serving four registers, and (c) multiple locations in RAM served by a single adder.

and the data source driving the bus is selected by setting the multiplexer inputs rather than by enabling and disabling tri-state buffers.

Designers should pay attention to these implementation details since a clean and neat design using tri-state buses may become resource consuming and slow upon conversion into multiplexers. Adding a data source in a tri-state bus system will not increase silicon resources other than the data source itself, while adding a data source in a multiplexer-based design will add an input port to the multiplexer, resulting in increased logic element usage in the FPGA. Implementing multiplexers with a large number of input ports usually needs multiple layers of logic elements, which may slow down the system performance if not appropriately pipelined.

Designers are recommended to review their interconnection requirements and to describe the interconnections explicitly in multiplexers rather than tri-state buses. It is often true that not all data destinations need to input data from all data sources. For example, assume there are six data sources U, V, W, X, Y, and Z and several data destinations A, B, C, D, etc. In the design, data destination A may only need to see data from U and V, B may only need to see data from V and W, and so on. In this case, it is preferable to use multiplexers with fewer input ports for data destination A, B, etc., rather than using a multiplexer with all six input ports. Very often, a data destination may only need to see data from one data source. In this case, simply connect the data source to the input of the data destination.

3.5   Remaining in control of the compilers

In FPGA design flow, after design entry, a compiler is invoked to convert the logical description into a physical configuration inside FPGA. For most general-purpose design jobs, compilers provided by vendors create reasonable results. However, given the wide variation of the FPGA applications, one cannot expect the compilers to always produce intended outcomes. In today's FPGA CAD tools, there are many switches, options that users can control. A suitable control on these options is a complicated topic that is beyond the scope of this book. In this section, we will discuss only a few simple issues and tips to use compilers.

3.5.1   Monitoring compiler reports on resource usage and operating frequency

It is recommended that designers frequently read the compilation reports to monitor the compiler operation and its outcomes.

Among the many items being reported, perhaps the resource usage and maximum operating frequency of the compiled project are the most interesting ones. When the resource usage is unusually higher than the hand estimate, poor design or compiler options are the usual causes. Excessive resource usage sometimes is coupled with a drop of maximum operating frequency.

3.5.2   Preventing useful logic from being synthesized away by the compiler

During the synthesis processes, the FPGA compilers convert the user's logic descriptions and simplify them for an optimal implementation. For most digital logic applications, the optimization provides fairly good results.

However, users may intentionally design circuits that are not optimal or with duplicated functional blocks for specific purposes, and compiler optimization is not required in these cases. For example, when a time-to-digit converter (TDC) is implemented using a carry chain to delay the input signal (see Chapter 6), the compiler may synthesize away the carry chain and directly connect the input to all the registers with minimum delay. Another example is in radiation tolerance applications (see Chapter 9) using the Triple Modular Redundant (TMR), when three identical functional blocks are implemented to process the same input data to correct possible errors caused by the single event upset effect. In the synthesis stage, the compilers may eliminate duplicated functional blocks.

It is possible to turn off certain optimization processes in the compilers, and sometimes it is even possible to compile the logic design in WYSIWYG (what you see is what you get) fashion. However, as the versions of the compilers are upgraded or as the design is ported to FPGA devices made in different companies, the exact definition of a particular optimization process may also change. Further, users may still want to apply general optimizations to most part of the design while preventing the compiler from synthesizing away useful logic in only a few spots.

A useful practice is to use the "variable-0's" or "variable-1's" to "cheat" the compilers. In the TDC implementation, an adder is used to implement a carry chain, and two numbers are input to the adder. Typically, a number with all bit set, that is, 1111…1111, and another number, 0000…000x, are chosen to feed the adder, where "x" is the input signal. When the input "x" is 0, the sum of the adder is 1111…1111; and when "x" becomes 1, bit-0, becomes 0 and a carry is sent to bit-1, resulting in it becoming 0 and sending a carry to bit-2, and so on. The propagation is recorded by the register array immediately following the adder, and a pattern such as 1110…0000 is captured at a clock edge; the position of the "10" transition represents the relative timing between the input signal and the clock edge. If the 1's or 0's used at the adder inputs are constants, the compiler will determine that the adder is unnecessary to calculate the final result and will eliminate it. To prevent the optimization from happening, variable-0's and variable-1's are used to construct the inputs to the adder. These variables are outputs of a counter that counts only during a short period of system initialization, and after initialization, these bits becomes constants. In this case, the compilers will not eliminate the adders.

A similar trick can be used in TMR for radiation-tolerant applications. The compilers usually identify "unnecessary" duplicated functional blocks by checking if the inputs of the functional blocks are identical. One may simply add variable 0's to the inputs feeding the three functional blocks so that the compiler will "see" that the three functional blocks are processing three different input values and will not eliminate them. Note that the three variable 0's must look different. Swapping some bits in the variable 0's will make them apparently different.

3.5.3   *Applying location constraints to help improve operating frequency*

In a pipeline structure, the propagation delay from the output of a register to the input of the register in the next stage determines the operating frequency. The propagation delay consists not only of the delay due to the combination logic functions, but also the routing between the two stages. In FPGA devices, routing resources with various connecting distances and propagation delays are provided. It is often true that the interconnections between logic elements physically close up are faster than the distant ones.

In an application with high operating frequency (>80% of maximum toggling frequency of the FPGA device) and relatively full usage of the logic elements (>70% of total), the compiler may start having difficulties in finding a layout of all the logic elements that meets all timing conditions. In this situation, the users may assign physical locations of the logic elements for the time-critical parts of the design to help the compiler fulfill all timing requirements.

The location constraints in FPGA design software are usually text based, and the following are examples from a design in an Altera FPGA device:

```
set_location_assignment LCFF_X3_Y5_N1  -to  TDC*:CHS0|*TCH0|QHa1
set_location_assignment LAB_X3_Y4  -to  TDC*:CHS0|*TCH0|QD1*
```

The first line assigns a particular register to the logic cell flip-flop (LCFF) N1 located in column 3 and row 5 in the device. The second line assigns several registers to the logic array block (LAB) at column 3 and row 4. Note that wildcards (*) are allowed so that one line can be used to assign several items, and one can simply use a wildcard as a shorthand for identifiers of items.

The constraints can be created using a text editor, but it is convenient to use a spreadsheet application such as MS Excel to manage to parameters such as X, Y, and N. The parameters and related text are concatenated together to form the assignment commands. The spreadsheets can be output as text files, and the text can be copy-pasted into the constraint file. Using spreadsheets, several hundreds lines of assignments can be easily handled. By combining with wildcards, several thousands of time-critical items can be placed, which should be sufficient for most applications.

3.6   *Guideline on pipeline staging*

It is known that breaking complex logic processes into smaller steps increases the system throughput, that is, increases the operating frequency of the system clock driving the pipeline. The pipeline operating frequency should be planned at the early design stage. An appropriately chosen pipeline operating frequency helps reduce the usage of precious FPGA silicon resources and thereby reduces system cost.

If the FPGA processes input data from other devices in the system, sometimes the pipeline operating frequency is chosen to be in a convenient ratio of the data-fetching rate. For example, if data are input at 50 M words/s, the operating frequency can be chosen as 200 MHz so that each processing pipeline can serve four input channels.

Another factor to be considered is that the operating frequencies of RAM blocks or multipliers in the FPGA are usually lower than that of logic elements. In the planning stage, it is a good practice to test these blocks first in a simple project before utilizing them in actual designs. Typically, the RAM blocks and multipliers can be configured with registers for both input and output ports to maximize their operating frequency, which is strongly recommended.

As we know, simple combinational logics are implemented with small lookup tables with typically four inputs in logic elements. When a combinational logic requires more inputs, logic elements are cascaded into multiple layers, resulting in longer propagation delays. When the pipeline operating frequency is chosen to satisfy the requirements for RAM blocks or multipliers, combinational logics consisting of logic elements will normally not become the bottleneck. Usually three to four layers of logic elements between pipeline stages will not impose a limit on operating frequency. However, care must be taken with elements using carry chains, such as adders. In an adder, the most significant bit may depend on the least significant bit that requires the signal to propagate through a long carry chain. Therefore, it is not recommended to cascade adders with other combinational logics, especially with other adders.

In summary, pipeline stages should be arranged for

- The input and output ports of the RAM blocks or multipliers
- The output of adders
- Combination logic longer than four layers of logic elements

In the FPGA devices, D flip-flops are designed in all logic elements, RAM blocks, and multipliers. Adding pipeline stages will not significantly increase silicon resource usage.

3.7   Using good libraries

Intellectual property (IP) cores, or other reusable code, are available for the FPGA designers. The quality varies over a very wide range. Before incorporating them into users' projects, it is recommended to evaluate them in a test project. Comparing resource usages of the compiled result and the hand estimate gives clues regarding the internal implementation and helps the designers to better understand the library items.

References

1. J. Wu et al., FPGA Curved Track Fitters and a Multiplierless Fitter Scheme, *IEEE Trans. Nucl. Sci.* vol. 55, no. 3, pp. 1791–1797, June 2008.
2. J. Wu et al., The Application of Tiny Triplet Finder (TTF) in BTeV Pixel Trigger, *IEEE Trans. Nucl. Sci.* vol. 53, no. 3, pp. 671–676, June 2006.

chapter four

Examples of an FPGA in daily design jobs

4.1 LED illumination

The LED is a convenient and low-power component suitable for indicating the low-level operational status of FPGA devices. Whenever a new board using an FPGA is designed, it is recommended to connect at least one LED to the FPGA. After the first new board is assembled and powered up, perhaps the most useful firmware to be downloaded into the FPGA is the one that makes the LED blink. When the LED starts blinking, it indicates that many important details have been designed correctly, such as power and ground pins for both the FPGA and configuration device, configuration mode setting, compiler software setting, etc. The LED blinking firmware in the FPGA design is as essential as the "hello world" program in the C programming language.

4.1.1 LED rhythm control

FPGA devices are often driven by clock signals with frequencies ranging from 10 to 100 MHz, while human eyes are much slower. Conventionally, resistors and capacitors on printed circuit boards are employed to create the necessary time constant for monostable circuits to drive LEDs. In the FPGA, however, it is more convenient to use multibit counters to produce signals in the frequency range of a few Hz, as shown in Figure 4.1a.

In the example of a 24-bit counter with 16 MHz clock input, it can be calculated that the toggling frequency of bit 23, the highest bit, is approximately 1 Hz. An LED connected to bit 23 will be turned on for 0.5 s and will be off for another 0.5 s.

One may enrich the rhythm of the LED flashing by adding some simple logic. For example, the LED output produced with the AND gate Q[23]. AND.Q[21] will create a double flash, 1/8 s on, 1/8 s off, 1/8 s on again and 5/8 s off until the next period, and the LED output produced with Q[23]. AND.Q[20] will create a strobe of four flashes, etc. Variations of the LED flashing rhythms can be used to indicate different operation modes of the FPGA, or simply to indicate the version of the firmware loaded into the FPGA.

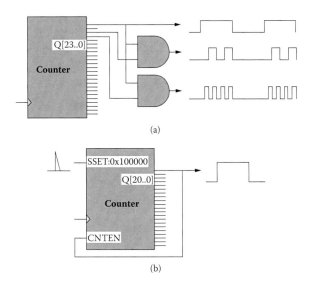

(a)

(b)

Figure 4.1 Counters for LED blinking: (a) repeating rhythm schemes, (b) short pulse display scheme.

In many applications, LEDs are used to indicate short pulses. A possible scheme is to use a counter with preset and count enable capabilities, as shown in Figure 4.1b, to stretch the short pulse to a length visible by human eyes. The counter has 21 bits and Q[20], the highest bit, is output to the LED. The signal Q[20] is also used as a count enable signal (CNTEN) and, after initialization of the FPGA, all bits are held to 0, and the counting is disabled. When a short pulse in sync with the clock arrives at the synchronized preset input SSET, the counter is preset to 0x100000, that is, Q[20] is set to 1, which allows the counter to start counting and illuminate the LED. The LED remains on while the counter is counting from 0x100000 to 0x1FFFFF, a total of 1048576 clock periods, or about 1/16 s if the input clock is 16 MHz, which should be visible to human eyes. When the counter reaches 0x1FFFFF, it rolls over to 0 after next clock cycle, and the counter returns to its initial state and stops counting.

This scheme operates in multiple hit pile-up fashion. If a narrow pulse follows the previous pulse before the counter finishes counting, the counter will be preset back to 0x100000, and the counting will restart. Therefore, two close-up pulses will join together to create a longer LED flashing. If necessary, it is possible to include additional logic so that the circuit can operate in different ways.

In fact, stretching an input pulse is a simple microsequence. A similar scheme using a counter with preset and count enable inputs will be discussed in Section 4.2.

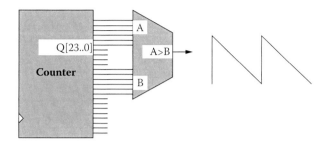

Figure 4.2 A scheme of changing LED brightness.

4.1.2   Variation of LED brightness

It is known that LED brightness varies with the current flowing through it. In the practical design of a printed circuit board, an LED is usually connected to an FPGA pin with a current-limiting resistor. The FPGA output pins normally only support logic levels, which yield constant brightness when the LED is on. However, when the flashing of an LED is sufficiently fast, its apparent brightness is lower and varies as the flashing duty cycle. A scheme of changing LED brightness is shown in Figure 4.2.

The duty cycle of the output is defined using a comparator with its B port connected to lower bits of a counter. Consider an example of a 6-bit comparator; at certain times, a value ranging from 0 to 63 is presented at the A port. The input value at the B port counting from 0 to 63 is compared with the value at the A port. During the 64 time periods for B port counting, the output is high only when B<A and, therefore, the duty cycle of the output pulse is A/64. The bigger the A port value, the brighter the LED appears.

Since the A port counts up slowly from 0 to 63, the brightness of the LED in this circuit increases gradually. When the A port rolls from 63 back to 0, the LED turns off and the brightness increases once again.

In fact, changing LED brightness by controlling the duty cycle is a useful scheme of digit-to-analog conversion (DAC) that will be discussed in a separate section. Using an analog low-pass filter, the output of the comparator becomes an analog voltage that is proportional to the duty cycle set at the A port.

4.1.3   Exponential drop of LED brightness

Human eyes have a very wide dynamic range with respect to brightness of objects. When the brightness of an LED varies linearly, the brightness change is felt too slowly at the high end and too rapidly at the low end. An exponential variation of brightness gives the human eye an impression of relative steadiness.

A common application of this circuit is to display a short internal pulse. Instead of generating a constant-brightness LED flash, a short pulse creates a sequence that causes the LED to turn on and then to dim down slowly. The exponential function can be generated simply with an accumulator in the circuit shown in Figure 4.3.

When the short pulse to be indicated is present, the accumulator is set to full range (0xffff for a 16-bit accumulator), which represents the highest brightness. A counter is used for two purposes: (1) to create the required duty cycle that is proportional to the brightness given at the A port of the comparator, and (2) to provide a timing tick from its carry output (CO) port that causes the accumulator to update.

The input data port (D) to the accumulator is a shifted version of its output (Q), enabled by the CO signal. When CO is 0, the input to the accumulator is 0, and the value stored in the accumulator remains unchanged. When the counter becomes full and rolls over to 0, the CO signal becomes 1 for one clock cycle, which causes the input of the accumulator to become a shifted version of Q (shifted right for five bits or $Q/32$ in the example given earlier). The value of the accumulator is reduced, and a new brightness value is presented to the A port of the comparator.

Assuming that the time interval between time ticks is Δt, the variation of the brightness Q satisfies the following equations:

$$Q(t+\Delta t)=Q(t)-aQ(t) \quad Q(t)=e^{bt} \quad e^{b\Delta t}=(1-a)$$

$$b=\frac{\ln(1-a)}{\Delta t}$$

(4.1)

According to these equations, the brightness drops exponentially, and the time constant is determined by the time tick interval Δt and constant a (1/32 in our example) that users can adjust to obtain the appropriate speed of LED dimming for the best visual effect.

It can be seen that relatively complex mathematic functions such as exponential sequences can be generated with very simple operations. In our example, not even a multiplier is used. In FPGA applications, many similar tricks are available, and designers are encouraged to use these resource-friendly tricks.

4.2   Simple sequence control with counters

In the FPGA design, an operation often needs multiple clock cycles to complete, which becomes a microsequence. Complex or reprogrammable

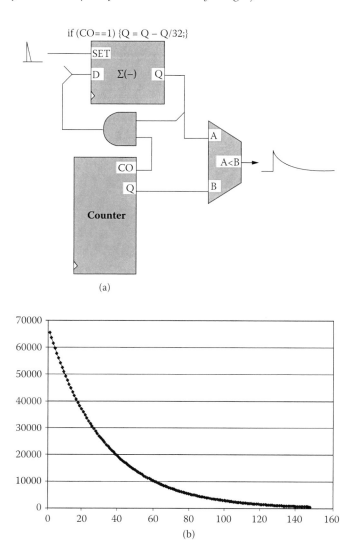

Figure 4.3 Exponential drop of LED brightness: (a) block diagram, (b) output of the accumulator as a function of time.

sequences are conducted using microsequencers or even microprocessors, while it is more convenient to conduct a broad range of fix sequences with simple counters.

4.2.1   Single-layer loops

Consider the partial design shown in Figure 4.4.

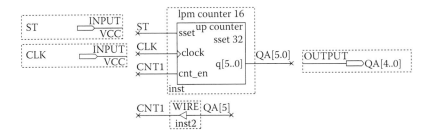

Figure 4.4 Single-loop sequencer implemented with a counter.

The counter has 6 bits, and its top bit is used to enable its counting. After power-up, all counter bits are initialized to 0, and the counting is disabled. When a pulse ST arrives synchronizing with the clock CLK, the counter is set to 32, that is, QA[5] = 1, and the counter is enabled to count, with QA[5..0] runs from 32 to 63, one step per clock cycle. It should be mentioned that the lower 5-bit integer, QA[4.0], runs from 0 to 31. After the 32nd clock edge, the counter rolls from 63 over to 0, and the counting sequence stops. The timing diagram is shown in Figure 4.5.

Similarly, controlled sequences of 0–63, 0–127, 0–255, 0–511, etc., can be generated with a counter of 7-, 8-, 9-, 10-bit counters, respectively, with the top bit being the counter enable signal.

Let us discuss several sequencing tricks through a useful example: addition of two arrays. Assume that two arrays, X and A, are stored in two memory blocks, each containing 256 elements. The two memory blocks must be addressed simultaneously in sequence, from 0 to 255, one element from each array per clock cycle. Once the elements from the two arrays are available at the outputs of the memory blocks, the two elements are added, and the sum is written back to the memory block storing array A. Note that the memory blocks are registered at both input and output ports. Therefore, there is a latency of two clock cycles from the address to the output data. In other words, when the address sequence 0, 1, 2, 3, etc., is presented at the address input, the memory content from address 0 is not immediately output until two clock cycles later, that is, when the address become 2. The contents in address 1, 2, 3 are output in the subsequent

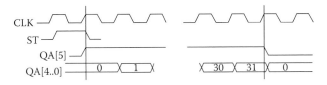

Figure 4.5 Timing diagram of a single-loop sequencer.

clock cycles, which are exactly two clock cycles after the input of the corresponding address. Taking memory latency into consideration, the address in our design for writing back the sum must be two clock cycles, later than the read address. Figure 4.6 shows a possible implementation.

The read address (RA) is implemented with a 9-bit counter and the top bit is used as count enable. The write address (WA) is an 8-bit counter with a synchronous clear input. The operating sequence is shown in the following timing diagram (Figure 4.7).

When a pulse ST arrives synchronizing with the clock CLK, the counter is set to 256, that is, CEA = 1. The counter is enabled to count, with RA[7..0] runs from 0 to FF, one step per clock cycle. Two clock cycles later, the contents stored in the memory blocks become available at the output ports MQX and MQA, one clock cycle per pair of elements. The elements of the two arrays are added and the sum, SAX, is sent to the input data port. The signal CEA is delayed by two clock cycles and becomes signal WE, the write enable signal. The counter for write address WA is usually held to 0 with the inverse of WE bring synchronous clear input. When the signal WE becomes 1, the counter for WA is allowed to count from 0 to FF, one clock per step. However, WA is always two clock cycles later than the corresponding RA, and therefore is aligned with the input data SAX.

4.2.2 Multilayer loops

We use a practical example to illustrate implementation of the multilayer sequences. Consider a complex sequence with two indices QAA[7..0] and QBA[7..0]. We need a two-layer nested loop followed by a single-layer loop. In the two-layer nested loop, QBA takes values from 0 to FF, 256 clock cycles per value, and for any QBA value QAA runs from 0 to FF, one clock cycle per step. Immediately following the end of the two-layer nested loop, the single-layer loop starts. In the single-layer loop, QBA counts from 0 to FF, one clock cycle per value. The index QAA is held to 0 for two clock cycles and starts to count from 0 to FF. At the end of the loop, QBA rolls over from FF to 0 and is held to 0, and two clock cycles later, QBB rolls over from FF to 0, and the whole sequence stops.

The sequence is implemented with three counters, two with 8 bits each and one with 2 bits. The sequence is conducted with an appropriate design of the control signals of the counters. The 2-bit counter can be synchronously set to 2, and the 8-bit counter can be synchronously cleared. Each counter has a count enable signal. The implementation is shown in Figure 4.8.

The sequence of the two-layer nested loop is shown in the timing diagram in Figure 4.9. After power-up of the FPGA, all counters are initialized and held to 0. When the start signal (ST) is seen at a leading edge of the

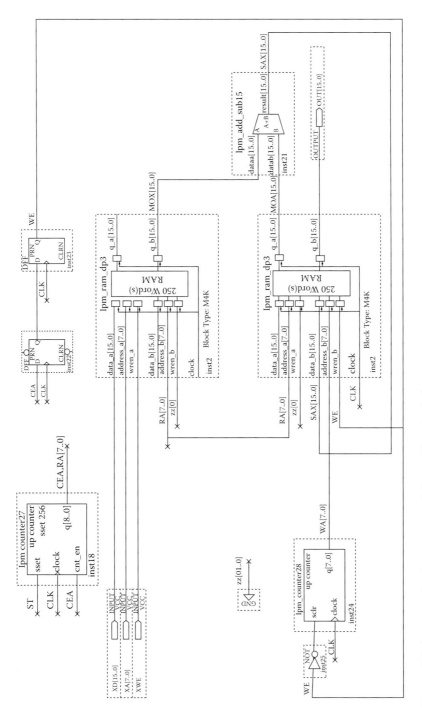

Figure 4.6 An example of memory updating with a single-loop sequencer.

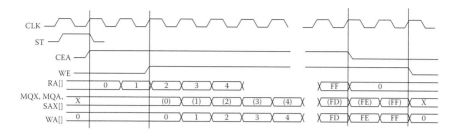

Figure 4.7 Timing diagram of the memory updating sequence.

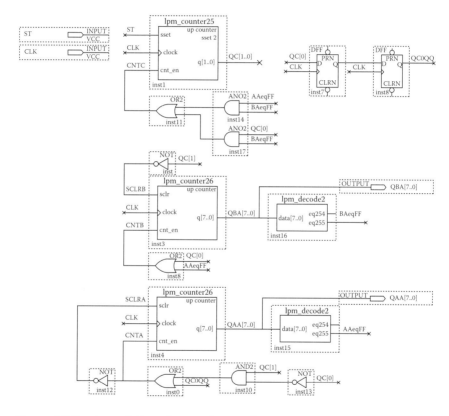

Figure 4.8 Implementation of a complex two-layer sequence.

CLK signal, the two-bit counter is set to 2, that is., QC[1] = 1 and QC[0] = 0. The counter for index QAA is enabled to count from 0 to FF, one count per clock cycle.

When QAA counts to FF, the signal AAeqFF becomes high for one clock cycle. The count enable signal CNTB derived from AAeqFF also becomes high for one clock cycle, allowing the counter of QBA to increase

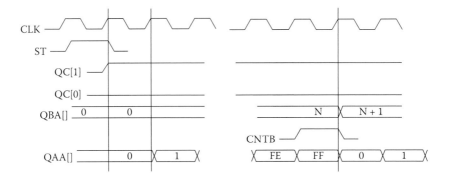

Figure 4.9 The two-layer nested loop timing diagram.

by 1. The QAA rolls over to 0 and continues to count up for another rotation. The whole sequence takes 256×256 clock cycles, with QAA repeating 256 times the counting from 0 to 255 and QBA growing from 0 to 255
with 256 clock cycles per step.

At the end of the two-layer nested loops, both QBA and QAA are FF,
as shown in Figure 4.10. This causes QC to increase by 1 (from 2 to 3)
and, also, the counters QBA and QAA roll over from FF to 0. Now, a new
looping condition starts. The count enable signal CNTB becomes constant
high because QC[0] = 1, which causes QBA to count up continuously from
0 to FF, one clock cycle per step. When QBA becomes FF, QC is enabled to
roll over from 3 to 0, and both QBA and QC are held to 0. At the start of
this single loop, QAA is first held to 0 for two clock cycles because signals
CNTA = 0 and SCLRA = 1. When the signal QC0QQ, the delayed version
of QC[0], becomes 1 after a two clock cycle delay, CNTA= 1 and SCRA =
0 again. The counter QAA now starts to count from 0 to 255, but it is two
clock cycles later than QBA. The counting of QAA stops when QC0QQ
becomes 0, which is again two clock cycles later than when QC rolls over
from 3 to 0. This completes the entire sequence and the system is ready for
the next start command, i.e., the ST signal pulse going high.

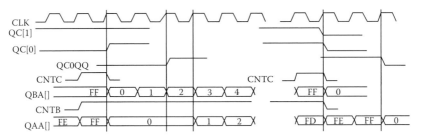

Figure 4.10 Timing diagram of the ending sequence.

4.3   Histogram booking

Histograms are useful tools for both instrumentation diagnosis and final physics analysis. Typically, histograms are booked offline in data acquisition computers, but in more and more practical tasks, online histogram booking is becoming necessary. For example, histograms can be booked to create online calibration lookup tables. Sometimes, the values for the histograms are only available inside the front-end FPGA and are not available in the data acquisition computers; in this case, the histogram can only be booked inside an FPGA. In the following text, we describe simple schemes for histogram booking and discuss a few issues regarding high-performance histogram-booking circuits.

4.3.1   Essential operations of histogram booking

When there are only a few bins in a histogram, it can be implemented using an array of counters, dedicating a counter for each bin. When the variable of an event falls into a given bin, the corresponding counter is enabled to increase the valuable by 1. After a sufficient number of events, a histogram booking is completed, and the values of the counters are output. This scheme can be used as a model to understand the online histogram facility, but it is not resource friendly. A counter contains a set of registers for data storage and an add-by-1 adder for incrementing the count. For each event, only one counter is enabled to increment and the adders of all the other counters are not used, which becomes a large resource waste when the number of bins in a histogram is large. In better implementation schemes, data storage of the bins and the add-by-1 adders are separated.

In our recommended scheme, histograms are implemented with random-access memories (RAMs), and in an FPGA, dual-port RAMs supporting read and write operations simultaneously are usually available. A simple histogram-booking circuit is shown in Figure 4.11.

The RAM is configured as a simple dual-port with a read port and a write port. The buses RA, WA, and D are read and write addresses and data to be written, respectively. The signal WE is the write enable signal. The input side for both read and write operations is registered, and the memory addressed by RA is output on bus Y. The value to be booked is first regulated into a bin number K. Sometimes, the input value is simply truncated into the bin number, keeping the higher bits and ignoring the lower bits. For example, if the input is a 12-bit value V[11..0] and a 256-bin histogram is to be booked, simply assign K[7..0] = V[11..4] and ignore V[3..0]. The bin number is accompanied by a data valid (DV) signal, which indicates that the bin number K is valid and the corresponding bin is to be accumulated.

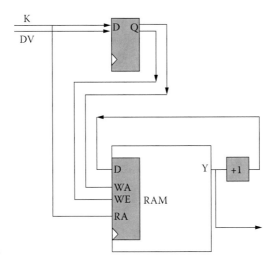

Figure 4.11 Simple histogram-booking circuits.

The RAM block is first initialized, with 0's being stored in all bins. The bin number K is used as the read address to read out the contents of the bin and, meanwhile, K and DV are temporarily stored in a set of D-type flip-flop registers after the first clock edge. Then, the content of the addressed bin is read out from Y, and it is increased by 1. At the same time, write address WA and write enable signal WE are presented at the write port. After the second clock edge, the value (Y+1) is written back into the addressed bin.

Histogram booking consists of three essential operations: (1) reading out the content of the addressed bin, (2) incrementing the content by 1, and (3) writing the incremented value back to the addressed bin.

This histogram-booking scheme fits the requirements of a broad range of applications, and it is able to accept a new event every clock cycle. However, data dependencies exist in its essential operations, and the data dependencies result in an operation restriction: the same bin shall not be hit in two consecutive clock cycles. If a bin is hit in two consecutive clock cycles, both read address RA and write address WA point to the same memory location, which may corrupt the result being read out at Y. This is similar to the read-after-write (RAW) hazard in microprocessors. In some FPGA families, writing to and reading from the same location is supported by dual-port memories, and the content being read out can be chosen to be "new," that is, the data to be written. In this case, the hazard is automatically resolved. However, it is always safer to satisfy the restriction by rearranging the input sequence so that the design can be ported among different FPGA families without causing a data dependency hazard.

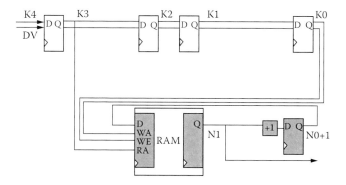

Figure 4.12 Pipelined histogram-booking circuits.

If a high operating frequency is required, the RAM block can be configured so that both input and output sides are registered. A pipeline can be arranged as shown in Figure 4.12, and in this scheme the adder is also registered to improve the operation throughput. This pipelined histogram-booking scheme allows a new hit input every clock cycle at a higher frequency. However, the RAW hazard due to data dependencies imposes a more severe restriction, namely, that the same bin shall not be hit within 4 clock cycles.

4.3.2   Histograms with fast booking capability

If in an application of the pipelined histogram scheme, the restriction that the same bin shall not be hit within 4 clock cycles cannot be satisfied, one may make improvements as shown in Figure 4.13 so that fast booking is supported regardless of the interval of rehit on the same bin.

The scheme shown in Figure 4.13a is similar to the data-forwarding scheme used to solve the RAW hazard in microprocessor design, while the scheme shown in Figure 4.13b can be called instruction merging, which is feasible only for histogram-booking functions.

The logic blocks marked with "&&==" are used to find two valid hits with identical bin numbers within 4 clock cycles. Once the rehit of a bin is discovered, appropriate processes are performed to avoid the RAW hazard.

In the instruction-merging scheme shown in Figure 4.13b, the bin number corresponding to the current output of the RAM is compared with bin numbers in the three later pipeline stages. If the current bin is rehit within 4 clock cycles and N (=0, 1, 2 or 3) valid hits with identical bin numbers are found, the output of the RAM is incremented by $(1 + N)$, taking the rehit into account. In the DV pipeline, the input to the next stage is forced to 0 if it is a rehit event so that the pipeline slot becomes invalid, to prevent the event from being recounted.

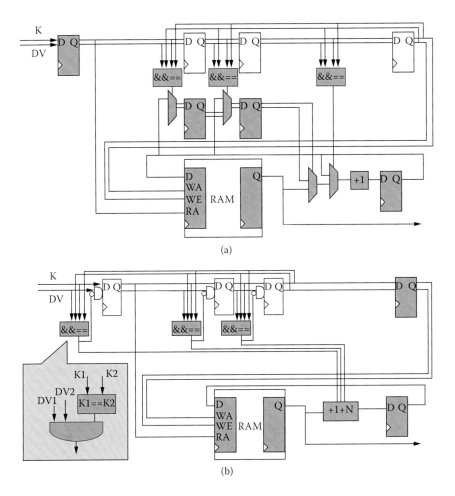

Figure 4.13 Fast booking histogram circuits: (a) the "data-forwarding" scheme, (b) the "instruction merging" scheme.

The operation of +1 in histogram booking can be considered as an "instruction." If the rehits are found and the output of the RAM is incremented by (1 + N), effectively these (1 + N) instructions are "merged" together and, hence, the scheme is named "instruction merging."

In the data-forwarding scheme shown in Figure 4.13a, the bin number corresponding to the registered result of the adder is compared with the bin numbers of the three later pipeline stages. If valid hits with identical bin number are found, the result of the adder register is selected and stored in the corresponding stage of a data pipeline (just above the RAM block in the diagram). In the later clock cycles, these selected data, rather than the output of the RAM, are used as input to the adder. (If the same bin is rehit in two consecutive clock cycles, the output of the adder register is directly chosen as the input of the adder.) In this case, the same location in the RAM may be read while being written, and the data read out can be corrupted. However, the data needed for the adder are "forwarded" in the data pipeline, and the output from the RAM is disregarded. A similar scheme is used in pipelined microprocessor design to avoid the RAW hazard.

4.3.3   Histograms with fast resetting capability

Unlike in flip-flop registers, a global reset signal is not supported in block RAMs. To reset the contents of a histogram, one needs to write 0 to N bins, which takes N clock cycles. For many applications, spending N clock cycles to clear the histogram between two sessions of booking is acceptable, but there are cases where a fast reset is needed. A possible scheme of a histogram with fast resetting capability is shown in Figure 4.14, which needs just one clock cycle to prepare for a new histogram-booking session.

The lower portion of the diagram is a regular pipelined histogram, and the fast resetting function is implemented in the top portion, consisting of a run counter (RC) and a RAM block called index RAM.

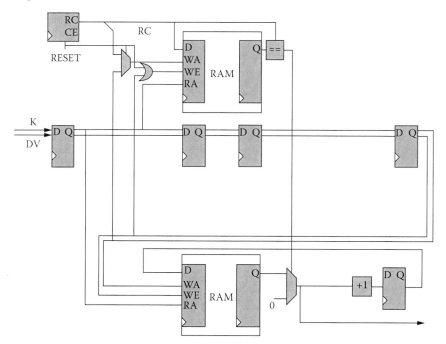

Figure 4.14 Histogram with fast resetting capability.

To explain the principle of operation better, consider a histogram containing 256 bins; therefore, the bin number K is an 8-bit integer. The index RAM has 256 bins, and the data width is 9-bit, matching the width of the run counter output RC. The run counter is enabled by the reset signal and, therefore, it increases by 1 for each histogram-booking session.

When a valid hit arrives, the bin number K becomes the read address for both index RAM and the histogram RAM. After the latency of two clock cycles of the RAMs, the contents of both the index RAM and the histogram bin are available at the output ports of the RAMs. The index is compared with RC, and the result of the comparison determines if the current bin is addressed the first time during the booking session or not.

If the index is different from the RC, the bin is first addressed in the session, and the content from the histogram RAM output is the data left from past sessions; therefore, 0 is selected for the input of the adder. The result from the adder, that is, 1, is written back to the bin to indicate that there has been 1 hit in the bin. At the same time, the same bin in the index RAM is written with the RC value, indicating that this bin has been addressed in the current booking session.

When a bin is addressed again, the output of the index RAM equals RC, indicating that the content stored in the histogram RAM is valid and should be selected as an input of the adder. The value of the histogram bin is then incremented by 1 and written back.

It can be seen that during the reset, the content of the histogram RAM is not cleared to 0. Instead, an index RAM is used to validate the histogram bins in each booking session. If the bin has never been addressed in a session, it is considered to be 0, although some old value from previous sessions may be stored in the RAM location.

There is a potential roll-over problem, however, given a run counter with a finite number of bits. A 9-bit run counter rolls over every 512 sessions, and it is not possible to distinguish the booking session S with sessions S+512, S+1024, etc. If not implemented correctly, a bin addressed in session 3, for example, will be mistakenly considered to have been addressed in session 515. To prevent the error caused by roll-over of the run counter, resetting logics are added to the write port of the index RAM.

During the reset cycle, the current value of RC is written into a bin in the index RAM. The bin to be written is chosen by rotation, and a convenient choice is simply the lower 8 bits of RC. This way, in any given session, all the bins in the index RAM are certain to be written to at least once in the past 256 sessions. At the beginning of any session S, the values stored in bins in the index RAM can only be one of 256 values from mod(S-1, 512) to mod(S-256, 512). Obviously, these 256 values will not be mistakenly interpreted as S in the 9-bit value.

It should be pointed out that fast resetting and fast booking features are two different requirements of online histograms. If both are needed in an application, the design methods described earlier can be combined together to create a circuit satisfying both requirements, but for simplicity, we will not cover this topic here.

4.4   Temperature digitization of TMP03/04 devices

The TMP03/TMP04 [1] is a monolithic temperature detector that generates a modulated serial digital output that varies in direct proportion to the temperature of the device. The output of the device is a square wave of approximately 35 Hz. The pulse high time T1 is approximately 10 ms and is less sensitive to temperature, while the pulse low time T2 varies from about 15 to 44 ms depending on the temperature.

The relationship between the temperature and the duty cycle can be written as

$$\text{Temperature (°C)} = 235 - 400*T1/T2 \qquad (4.2)$$

The time intervals T1 and T2 can be measured fairly easily with counters. However, a division, a multiplication, and a subtraction are necessary to calculate the temperature. Perhaps the simplest implementation is to keep two counters for the two time intervals and to read the values of the two counters into a microprocessor. The arithmetic operations necessary to calculate the temperature can be done in the microprocessor.

If the temperature is to be calculated in the FPGA, not only time interval measurement but also arithmetic operations must be carefully planned. Yes, there are divider and multiplier macro function libraries, but they are normally for many fast computations and are resource heavy. In temperature measurement, many clock cycles are to be spent counting the lengths of the time intervals T1 and T2, so smaller (although slower) arithmetic functions would be ideal.

Here, we discuss a counter-based circuit as shown in Figure 4.15. In this scheme, the arithmetic operations are performed during counting time intervals T2 and T1. This may not be the best circuit for all applications, but we will use it as an example to illustrate alternative approaches when arithmetic operations are needed in an FPGA.

The temperature output from the foregoing circuit is an integer with the LSB of 1/16 (°C). The formula of temperature for this case can be written as

$$T16C \ (1/16°C) = 16*235 - 16*400*T1/T2 = 3760 - 25*T1/(T2/256) =$$
$$3760 - 25*T1/(T2N>>8) \qquad (4.3)$$

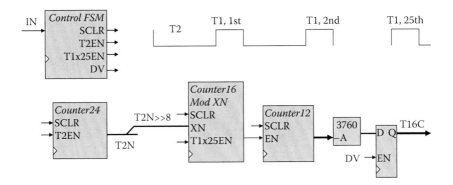

Figure 4.15 The temperature digitization circuit.

The modulated square wave IN signal is sent to the control finite state machine (Control FSM). The FSM first generates a clear signal SCLR to reset all counters. Then, during the first IN = low time interval, signal T2EN is active, which enables a 24-bit counter (Counter24) to count the length of T2; the result is a 24-bit integer T2N. With a 50 MHz system clock, T2N may have 22 nonzero bits. The shift right operation is simply ignoring the lower 8 bits and taking only the top 16 bits.

The next 16-bit counter (Counter16) uses the (T2N>>8) input as mod divisor. It outputs a pulse every (T2N>>8) clock cycles. The output then enables a 12-bit counter (Counter12) to increase by 1 each time. The Counter16 is enabled by the signal T1x25EN, which is active when the input IN becomes high, for 25 such intervals. The output of Counter12 then represents the integer value 25*T1/(T2N>>8).

After 25 T1 intervals are counted, the FSM generates a data valid signal DV, which enables the output register to update the new temperature value. The output of Counter12 flows through an adder that subtracts 25*T1/(T2N>>8) from a constant 3760. The registered result T16C is an integer with a unit of 1/16 (°C). It is refreshed every 25 T1 intervals or about 0.7 s, which is sufficient for most applications.

The division in this scheme is absorbed in the counting of Counter16 and Counter12. The multiplication of 16*400 is split into factors 25 and 256 and is done by counting 25 T1 intervals and by bit-shifting on T2N, respectively.

4.5   Silicon serial number (DS2401) readout

The DS2401 [2] is a 1-wire bus device that provides a unique 64-bit registration number (which contains 8-bit family code + 48-bit serial number + 8-bit CRC tester). The 1-Wire bus devices have only two connected pins, a ground and a data (DQ). The DQ pin is pulled up through a resister, and all I/O accesses are through the DQ pin (probably the device obtains power from the DQ pin also).

It takes three operations to read out the number from the DS2401 device. During the first step, the master FPGA or microprocessor drives the DQ line low for more than 480 μs to initialize the DS2401 device.

In the second step, the master writes an 8-bit command 0Fh (or 33h) to the DS2401, indicating that the device is to be read. To write a bit with value 1, the master drives the DQ wire low for a short period of time (1 to 15 μs) and releases it. The DS2401 senses the DQ wire level to be 1 in the sampling window since DQ is pulled up by the resistor. To write a bit with value 0, the master drives the DQ wire low for a longer period of time (60 to 120 μs) that covers the sampling window of the DS2401, resulting in a 0 level being detected. Once the 8-bit sequence 1,1,1,1,0,0,0,0 (0Fh, LSB first) or 1,1,0,0,1,1,0,0 (33h, LSB first) is written into the DS2401, it is ready to output its content.

The master reads out 64 bits in the third step. To read each bit, the master drives the DQ wire low for a short time and releases it. The master then senses the DQ wire to determine if the output from the DS2401 is low or high. The reading sequence is also LSB first.

There could be many schemes to generate the necessary pulsing sequence to read out the serial number. A circuit using ROM for sequencing and dual-port RAM for deserialization is shown in Figure 4.16.

In the foregoing example, a 45 MHz clock drives a 26-bit counter continuously to flash an LED as a clock indicator. The middle bits of the counter C[18..9] are taken to generate the sequence for serial number readout. The lower 3 bits C[11..9] are defined as "pulse time" PT[2..0] = 0 to 7, with 11.4 μs each step. Each set of 8 pulse time steps is grouped as a bit time slot, which is indexed by 3 bits C[14..12]. The higher 4 bits of the counter C[18..15] are used as the byte index. Reading out 64 bits from DS2401 takes 8 bytes. Before reading, the device must be initialized and fed with a read command. So 16 bytes are reserved in the byte index.

A ROM is used to generate the pulsing sequence on signals DQ, the 1-Wire bus pin, and signal WE, the write enable input of the RAM. The ROM contains 2048 bits, which can be implemented with a block memory in FPGA. In fact, there are many unchanging and repeating sections in the sequence, so that the whole ROM can also be composed with a few small sub-ROM pieces, which can be implemented with the LUT resources in the logic elements.

The initiation uses the entire time at byte time 5, that is, C[18..15] = 5, which is 728 μs long, followed by another 728 μs device recover time during byte time 6. The byte time 7 is used to feed the read command to the DS2401, which contains 4 Write-1 and 4 Write-0 bit time slots. The read sequence uses byte times 8–15, during which time the WE becomes active at PT = 2 in each bit time slot so that the data from the DS2401 can be written into the RAM.

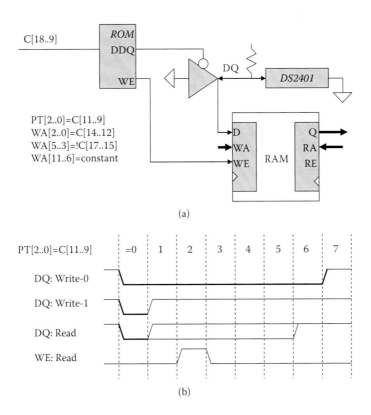

Figure 4.16 The silicon serial number readout circuit: (a) block diagram, (b) timing diagram of read and write sequences.

The write and read ports of the RAM are 1 bit and 16 bit wide, respectively. Data are written into the RAM bit by bit and read out as a 16-bit word. The write addresses are simply derived from the counter bits C[17..12]. Note that the bit, byte, or word order can be reversed to fit requests of different applications. For example, to maintain bit order and reverse the order of all 8 bytes, let WA[2..0]=C[14..12], but let WA[5..3]=!C[17..15].

The RAM block in our example has 4096 memory bits, while the silicon serial number uses only 64. The higher bits of the WA are set as a constant so that the 64-bit silicon serial number is placed in the predefined location.

Clearly, it is also possible to use a shift register to convert the serial stream into parallel words and, indeed, it is a more classical approach. In our example, the content of the RAM block is initialized into ASCII characters so that it is read out as an ID block, and the silicon serial number of the module is embedded into the ID block automatically. If using a shift register, the ID block and the shift register are two data sources, and they have to be merged together into the readout data path with a multiplexer. Also, other miscellaneous serial data such as temperature and power supply voltages can also be merged into the RAM block that provides a one-stop ID/status block for the system diagnostic process.

References

1. Analog Devices, Inc., TMP03 / TMP04 Serial Digital Output Thermometers, 2002, available via: http://www.analog.com/.
2. Maxim Integrated Products, DS2401 Silicon Serial Number, 2006, available via: http:// www.maxim-ic.com/.

chapter five

The ADC + FPGA structure

In a broad range of practical applications, input signals are digitized with an analog-to-digit converter (ADC), and the digital data are immediately sent to an FPGA for processing. Sophisticated signal processing functions such as filtering, signal feature extraction, noise elimination, etc., can be conveniently performed digitally in an FPGA, or in microprocessors when the digital data are transported to later stages. Compared with analog approaches, digital signal processing is easier to implement, and more flexible for algorithm changing and parameter adjustment.

5.1   Preparing signals for the ADC

Digital signal processing can extract useful information only if the information can be represented as digital data and is separable from noise. Therefore, for each system, several issues must be addressed in the analog domain before the signal is digitized in the ADC. It is recommended that designers at least review the following two requirements to prepare input signals for ADC: antialias low-pass filtering and dithering.

5.1.1   Antialiasing low-pass filtering

When a continuous signal is sampled with a train of impulses at equal time intervals, a phenomenon called *aliasing* may occur, causing different frequency components in the signal to become indistinguishable. For example, if the sampling frequency is f_s, the signal components with frequencies nf_s will appear as DC offsets in the sampled sequence. In general, higher-frequency components will contribute to the sampled sequence as if they are low-frequency components.

According to sampling theorem [1], a continuous signal can be correctly reproduced with the sampled sequence only if the continuous signal is band limited. The cutoff frequency of the band-limiting signal must be less than half of the sampling frequency, that is, $f < f_s/2$.

An appropriate low-pass filter must be designed to limit input signal bandwidth to be digitized by the ADC. In high-energy physics and the nuclear physics community, the low-pass filters before the ADC are usually called *shapers*. In commissioning a practical system with the ADC+FPGA structure, various ADC sampling rates may be chosen for optimizing system performance. Therefore, in the design stage, the shaper parameters should be chosen to meet the band-limiting requirement for the lowest ADC sampling rate. For example, if in a system 5, 10, and 20 MHz ADC sampling rates are anticipated, the shaper cutoff frequency must satisfy the sampling theorem condition for the 5 MHz case; that is, the cutoff frequency must be <2.5 MHz.

On the other hand, if the cutoff frequency of a shaper has been fixed and a lower final sampling rate is needed, one should not simply reduce the ADC rate since that will violate the sampling theorem condition causing aliasing. If adjusting the shaper parameter is inconvenient, a correct approach is to use a higher ADC sampling rate that meets the condition required by the sampling theorem and to use a process called *decimation* to reduce the final data rate digitally inside the FPGA. Consider the foregoing system: if a final data sampling rate of 1 MHz is sufficient, the designers may adjust the cutoff frequency of the shaper to <0.5 MHz and run the ADC at 1 MHz. If the designers choose not to adjust shaper parameters due to inconvenience, they may run the ADC at a higher sampling rate, say, 8 MHz, and then reduce the data rate down to 1 MHz inside the FPGA, using the decimation process.

5.1.2   Dithering

Contrary to general intuition, under some conditions, adding noise to the input of the ADC, a procedure called *dithering*, may help improve measurement precision. In an ideal ADC, each ADC output code represents a range of input voltages, and the ADC output code will remain unchanged until the input voltage passes beyond the upper limit or lower limit of the range. Therefore, in a very quiet system with a slow-changing input voltage, the ADC output will stick to an identical code for many samples. With a series of identical ADC codes, there is no information on the variation of the input voltage within the range corresponding the ADC code. In other words, the best measurement precision for a quiet ADC input is the least significant bit (LSB) of the ADC.

Adding small noises to the ADC input can help achieve a measurement precision finer than the ADC LSB. An example is shown in Figure 5.1.

In this example, the input voltage changes slowly in a range corresponding to the ADC codes 52 and 53, and when the ADC is ideal, the output code changes from 52 to 53 if the input voltage passes through a threshold corresponding to the 52.5 ADC count. In the quiet system shown in Figure 5.1a, the output of the ADC is first a series of identical codes, 52, when the input is below the threshold, then becomes 53 for multiple samples when the input passes above the threshold, and then returns to 52 after the input returns below the threshold. From the ADC output, it is impossible to detect a variation of input finer than the LSB.

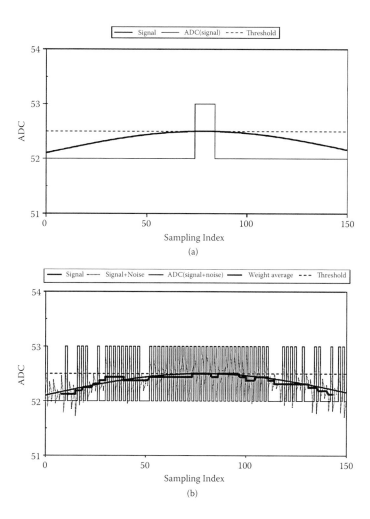

Figure 5.1 An example of the dithering procedure: (a) a clean signal and the ADC response, (b) the signal with noise, the ADC outputs, and the sliding averages of the ADC outputs.

If a small amount of white noise is added to the input of the ADC as shown in Figure 5.1b, there is a probability that the sum will pass above the threshold, even when the voltage to be measured is below the threshold. The ADC output will jump between codes 52 and 53. The information of the input voltage level within an LSB precision is carried into the sequence of ADC codes. Sliding averages (thick solid line) of the ADC outputs are calculated, and it can be seen that they trace the variation of the original signal with precision finer than the ADC LSB.

Note that if the precision of ADC LSB is sufficient for a certain application, explicit dithering is not needed.

To add white noise to the ADC input, one may use on-board noise generation circuits. But in most cases, the intrinsic noise from previous analog stages is large enough. Also, the ADC device itself may create a large noise that may be fed into its inputs.

The recommendation to designers is that the noise level at the ADC inputs should be carefully reviewed and tested. The noise level should be reduced if it is too high, but it is unnecessary to fully eliminate random noise using high-cost components since finer measurement precision can be achieved through the dithering procedure when a small amount of noise is present.

5.2   Topics on averages

In daily life, we calculate the average of several numbers to get something "better," more "precise," or more "credible." Average calculation is a very useful building block in an FPGA and it is the basis of other digital processing such as filtering. Averaging in an FPGA is similar to its counterpart in software in terms of general principles. Due to flexibility of the data format, the implementation in an FPGA is actually simpler than in software. We will discuss topics related to the averaging process in this section.

5.2.1   From sum to average

It is known that to calculate an average, data points are summed, and the result is divided by the number of data points. To avoid division, which is harder to implement in an FPGA, choose the number of data points to be the 2^K numbers, such as 2, 4, 8, 16, 32, 64, 128, 256, etc. When the 2^K numbers are summed together, the sum becomes the average automatically, with the higher bits being the integer portion and the lower K bits being the fractional portion of the average. Unlike in a software implementation, in which bit alignment has to be done explicitly, in FPGA the bit alignment is done implicitly while wiring the sum result to the later stages.

Consider an example of four 8-bit integers A, B, C, and D. When they are added together, the word length of the sum S = (A+B+C+D) becomes 10 bits, and the higher 8 bits and the lower 2 bits are integer and fractional portions, respectively, of the average.

5.2.2   Gain on measurement precision

When 2^K data points are summed together, the word length of the result becomes K bits longer, which is the fractional portion of the average. However, not all of these K bits are effective.

Consider that N data points represent an expected value and their errors have the same standard deviation σ_1. If the errors are independent of each other, the average of these N data points will reduce the error with standard deviation $\sigma = \sigma_1/\mathrm{sqrt}(N)$.

We can loosely view the limited precision of each data point (the truncating error) as one of the various error sources. After summing up 2^K data points, a factor of $1/\mathrm{sqrt}(2^K) = 1/2^{K/2}$ reduction of this error contribution is anticipated (with a few more assumptions regarding the nature of the error distribution). This reflects a gain of precision of $K/2$ bits in the fractional portion.

For example, if 64 data points are summed together, the word length of the result is 6 bits longer. However, at the most one can rely on the higher 3 bits, and the lower 3 bits should be ignored.

A problem must be pointed out here: the truncating error is usually not visible without other measurement error sources. In this situation, there will be no gain in measurement precision. This is why dithering (described earlier) is needed to achieve a measurement precision finer than the ADC LSB.

For example, if a voltage of 12.3 V is measured 64 times but each measured result is truncated to 1 V precision, we will get 64 identical measurements of 12 V. From these identical measurements, it is impossible to know that the original voltage is 12.3 V. But if some errors are added so that some of the measurements become 13 or 11 V, it becomes possible, loosely speaking, to achieve a measurement precision as good as 1/8 V with an average of 64 measurements.

5.2.3   Weighted average

In a regular average, all data points are treated equally, which yields a minimal standard deviation when all the data points have the same measurement precision and represent a constant value. In some applications, however, some data points are either more accurate or known to be close to the value to be measured, so that these data points are given higher weight factors.

Calculating weighted averages is a special case of finding inner products. In an FPGA, weighted averages can be implemented as inner products, and an example of calculating three weighted averages is shown in Figure 5.2.

The input data and corresponding coefficients (or weight factors) are sent to multipliers in sequence. Their products are summed up in the accumulators. Once all data points and the coefficients are fetched, the weighted averages become available in the accumulator, ready to be output to later stages.

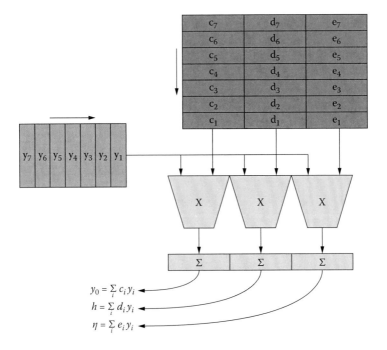

Figure 5.2 Weighted averages.

An appropriate definition of the bits for input data, weight factors, and output results is necessary to understand the normalization. In a typical FPGA, a multiplier with N-bit inputs has a $2N$-bit output, which sometimes is referred as an integer multiplier, since multiplying two N-bit integer yields a $2N$-bit product.

Without losing generality, let us assume the input data are 8-bit integers and the coefficients are 8-bit full fractional values; that is, the MSB of the coefficients is 0.5, and when all bits of a coefficient are set, its value is $1 - 2^{(-8)}$. The normalization condition requires that the sum of all coefficients be 1 (or 256 if the coefficients are viewed as integers). In the output of the accumulator, which is a 16-bit number, the higher 8 bits represent the integer portion of the weighted average and the lower bits, the fractional portion. The number of effective bits in the fractional portion depends on the actual values of the coefficients and usually is fewer than 4 bits. Only when all coefficients are identical, that is, when the average becomes the unweighted average, does the number of effective bits become 4, if there exist other noise sources.

5.2.4   *Exponentially weighted average*

Multipliers are usually needed for weighted average implementation. There is a simple yet very useful weighted average that can be

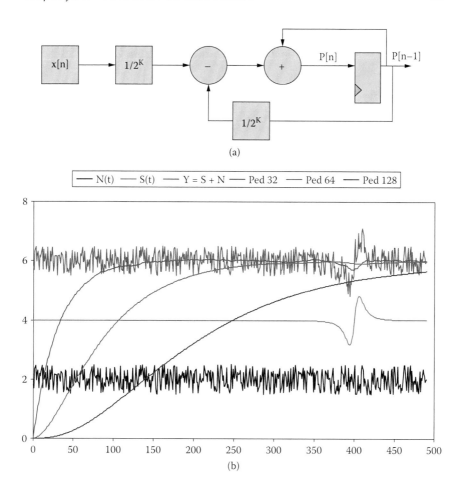

(a)

(b)

Figure 5.3 Exponentially weighted average: (a) block diagram (b) tracking process of the weighted averages with different time constants.

implemented without multipliers. In this weighted average, the weighting factor sequence is an infinite series with the highest weight for the current input data point and exponentially reduced weights for earlier points. In an FPGA, the weighted average can be implemented simply using an accumulator and an adder, as shown in Figure 5.3.

The input data is a sequence given in $x[n]$, and the output sequence $P[n]$ is given in the following differences equation:

$$P[n+1]=P[n]+((x[n]-P[n])>>K) \tag{5.1}$$

The bit shifting by K operation is equivalent to a division of a power-of-2 number, which determines the time constant of the exponential sequence. In Figure 5.3b, the input sequence $x[n] = N(t) + S(t)$, which is the sum of the noise and the signal, and Ped32, Ped64, and Ped128 representing the exponentially weighted average of input sequence with $2^K = 32$, 64, and 128, respectively, are shown. Full analysis of the difference equation is omitted here, and only a few special cases are inspected.

It can be seen that the value stored in the accumulator, $P[n]$, is primarily kept unchanged, with only a small correction if the difference $(x[n]–P[n])$ is nonzero. Clearly, if $P[n]$ has reached the average of $x[n]$ and $x[n]$ remains unchanged, the output will remain unchanged.

The user can choose value K for a different time constant so that the performance meets the requirement of the application. It can be seen from the foregoing that when the time constant is small, the output tracks to the input promptly but the curve is less smooth. For a longer time constant, the output is smoother, but it takes a longer time for the output to follow the change of the average. With the bit-shifting operation, only a few time constants are available. If additional time constants are needed, more adders or even multipliers can be used to implement the multiplication operation.

The exponentially weighted average is suitable for applications involving tracking of the relatively slow change of average values, such as pedestals from ADC inputs. The average calculator given here needs no initialization and no termination of the accumulation sequence as regular average calculators do. After a sufficiently long time, the output of the average calculator converges to the average and becomes available at all times.

5.3   Simple digital filters

As in the analog domain, digital filters are important building blocks for signal processing functions. Digital filters have various frequency responses such as low-pass, high-pass, band-pass, and band-stop types. The basis of digital filter design is low-pass filters, for the following two reasons. First, in a large range of applications, signals are in the lower-frequency region and noise is in high frequencies, so the low-pass filter is appropriate. Second, all other frequency responses can be derived from low-pass filters via appropriate transforms and combinations.

Generally speaking, digital filters are implemented as inner products of input data arrays, and the coefficient arrays and multipliers are needed. In this section, we do not intend to cover generic digital filter design issues but will concentrate on two simple yet very useful low-pass filters that do not use multipliers.

5.3.1   Sliding sum and sliding average

The sliding sum or sliding average of a sample series represents a moving trend of the input data. At a given time with index n, L input points x_i before n ($i = n - L - 1$ to n) are summed up:

$$s_n = \sum_{i=n-L-1}^{n} x_i$$

$$(5.2)$$

If L is a power-of-2 number, the sliding sum automatically becomes the sliding average. The series s_n and x_n have the same sampling rate; that is, for any new data point, a new sliding sum can be constructed.

   If the sum length L is large, directly implementing the sliding sum will need a large amount of computation (L add operations) that would consume a large amount of silicon resource in an FPGA. Silicon resource usage can be significantly reduced with the recursive implementation approach. Consider two adjacent sliding sums: it can be observed that most of data points in the two sums are identical. For the new sliding sum, the block of data points simply slides by one point, that is, adding a new data point and deleting an old data point as given in the following equation:

$$s_n = s_{n-1} + x_n - x_{n-L}$$

$$(5.3)$$

There are only two operations, an addition and a subtraction in the recursive implementation, no matter how large the sum length L is. Clearly, the amount of computation and therefore the FPGA resources needed will be much less. A block diagram of the sliding sum implementation in FPGA is shown in Figure 5.4.

   At any given time, the input data point is used to construct the sliding sum and meanwhile is stored in sequence in a pipeline (typically implemented using internal RAM in FPGA). Each time, the old data point taken from the end of the pipeline is subtracted from the old sliding sum, and the new input data is added to create the new sliding sum.

   At the start-up, when fewer data points than the sum length are input, the tail being subtracted should be zero. However, the content of the memory buffer may not be zero since block memories usually cannot be reset through a global command. One may certainly initialize the buffer by writing zeros into it, but the process takes a long time. A better way is to use an AND gate at the data path of the tail. The AND gate first suppresses the tail to zero before the number of input data points reaches the sum length. After the number of input data points becomes greater than the sum length, the AND gate allows the tail data to pass though.

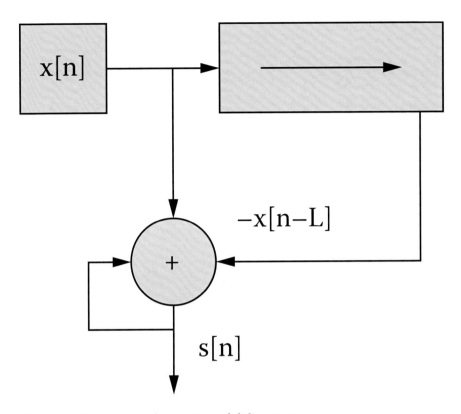

Figure 5.4 Recursive implementation of sliding sum.

5.3.2   The CIC-1 and CIC-2 filters

The cascaded integrator-comb (CIC) digital filter [2] of order N contains N cascaded stages. Each stage is a sliding average filter, which is a CIC filter of order 1. The sliding sum can be viewed as a CIC filter of order 1 with unnormalized gain. CIC filters of order 2 can be viewed as the sliding sum of the sliding sum of the raw data.

The frequency response shape of the CIC-2 filter is $sinc^2(x)$, in which the zeros become the second-order ones that provide deeper attenuation of the noise peaks even though the peaks are not precisely aligned with the zeros. The side lobes also become lower.

In the practical firmware, the CIC-2 filters are implemented recursively as shown in Figure 5.5.

Both diagrams in Figures 5.5a and 5.5b are valid CIC-2 sum implementations, and the resource usages are comparable. However, with the diagram shown in Figure 5.5b, we can avoid adding a separate storage for $s[n]$, given that a record of raw measurement points $x[j]$ is usually available.

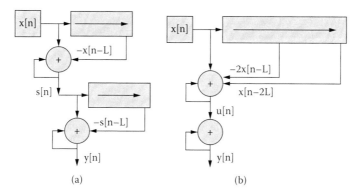

(a)   (b)

Figure 5.5 Recursive implementation of CIC-2 filters: (a) the CIC-2 filter implemented as cascaded CIC-1 filters (b) the implementation scheme using single delay line.

The formula for the CIC-2 sum altered for the diagram in Figure 5.5b contains two recursive accumulations:

$$u[n] = u[n-1] + x[n] - 2x[n-L] + x[n-2L]$$

$$y[n] = y[n-1] + u[n]$$

$$(5.4)$$

In this way, the only additional storage is the intermediate value $u[n]$, which takes only one memory space, and no additional long record of intermediate values needs to be stored.

5.4   Simple data compression schemes

With the availability of fast ADC and switch capacitor array ASIC devices, waveform digitization becomes possible in a broad range of applications. The shapes of the signal pulses can carry much richer information than the traditional charge integration scheme for resolving pileup in high rate systems or for precise timing measurement. An unavoidable challenge is that waveform digitization produces a large volume of data at the rate of one word per sampling point, creating difficulties for transporting and storing the data. In this section, we discuss several possible simple schemes of compressing waveform data.

5.4.1   Decimation and the decimation filters

As mentioned earlier, in an ADC + an FPGA structure, the ADC may provide oversampled data; that is, the ADC sampling rate can be many times higher than required signal bandwidth. In this situation, the FPGA can produce data at a lower sampling rate to reduce the data output rate and storage requirement. This process is called *decimation*. For example, if the ADC is sampled at 8 MHz and only 2 MHz data are needed to output from FPGA, a factor-of-4 decimation process is performed in the FPGA.

It should be pointed out that decimation is not simply "throwing away" unwanted data points. In the foregoing example, one output is needed for every 4 data samples; keeping one sample, and throwing away the other 3 is not an acceptable scheme. Intuitively, many designers would think that summing up 4 data samples and outputting the average should be good enough, but actually it is insufficient.

The decimation process consists of two essential steps: (1) antialias low-pass digital filtering and (2) downsampling. In the antialiasing filter, it is crucial to fulfill the requirement of the sampling theorem, that is, the cutoff frequency of the filter must be less than half of the new sampling frequency, $f < f_s/2$.

Averaging can be viewed as a low-pass filtering process, and it can be used as the decimation filter. However, care must be taken to review the sampling theorem requirement. In a sliding average with N samples, the first zero in the frequency response is at f_{s1}/N, where f_{s1} is the original sampling frequency (before decimation). In the example of decimation from 8 to 2 MHz, the required cutoff frequency after filtering is 1 MHz. If the first zero of the low-pass filter is viewed as the cutoff frequency, 8 samples, and not just 4, must be averaged, as shown in Figure 5.6. Note that there is an overlap of 4 data samples between the two adjacent decimation outputs.

As mentioned earlier, the averaging process is a CIC-1 filter, and there are higher-order CIC digital filters. Higher-order CIC filters such as the CIC-2 filter can also be used as a decimation filter. But again, the requirement of the sampling theorem on the cutoff frequency must be carefully fulfilled.

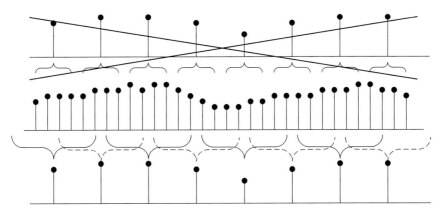

Figure 5.6 Averages as decimation filters.

5.4.2   The Huffman coding scheme

A common property of many waveforms is that the voltage levels between two sampling points do not differ by a large amount. In other words, if the value of a data sample is known, the values of subsequent data samples are likely to be near the previous one. Taking advantage of this property, it is possible to reduce the data volume without losing useful information. A scheme that can be easily implemented in FPGA is a subset of the Huffman coding scheme.

For a sequence of values of data samples $U(n)$, a difference is first produced between two adjacent values: $U(n)-U(n-1)$. The values of the difference are concentrated primarily around 0, +1, and −1 and larger differences may exist, but with fewer probabilities (Figure 5.7).

As shown in Figure 5.7, the values are assigned with predefined codes of variable lengths according to a coding table. Values with higher probabilities are assigned to the shortest codes, and less probable values are assigned to the longer codes. The codes are packed together to form 16-bit words. If the difference is >+3 or <−3, or if the data point is the first sample in a sequence, a full 16-bit data word with raw ADC value is used.

This coding scheme has been tested in a set of typical waveforms from a liquid argon time projection chamber, and a compression ratio of 10 was achieved. It can be seen that in this coding scheme, the theoretical maximum compression ratio is 15, which corresponds to the case where all raw ADC values are identical so that all differences are 0's. It is possible to improve the coding efficiency still further if the raw measurement values have a very slow variation and small noise so that many differences are 0's. For example, four adjacent 0's, 0000, can be assigned to the shortest code with one bit, and the maximum theoretical compression ratio can be 60.

Obviously, the Huffman coding scheme given here is lossless; that is, all raw data can be restored from the compressed codes.

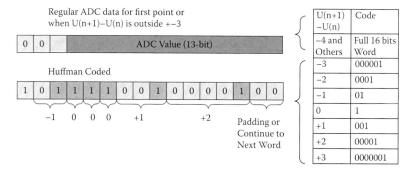

Figure 5.7 Huffman coding for waveform digitization.

The block for Huffman coding has been test designed, compiled, and simulated with 250 MHz operating frequency in an Altera Cyclone III FPGA device. The Huffman coding block first calculates the difference between the current and the previous data points and then finds the Huffman code according to the value of the difference through a coding table. The Huffman codes of several data samples are concatenated together, and a data valid signal is generated each time a 16-bit data word is filled up. The silicon resource usage of this block is around 245 logic elements, which is a small fraction in a typical low-cost device.

5.4.3   Noise sensitivity of Huffman coding

Since the Huffman coding scheme described earlier is based on the differences of raw data samples, naturally it is very sensitive to high-frequency noise. Care must be taken of the analog bandwidth of the shaper before the ADC starting from outside of the FPGA. Inside the FPGA, appropriate digital filtering must be applied to fulfill the requirement of the sampling theorem.

Several waveforms with an original sampling rate of 5 MHz are decimated down to 1 MHz, and the data compression ratios for different decimation filters are compared in Figure 5.8.

The original 5 MHz data samples are given in the leftmost column with compression ratios of about 10. In the second column, the compression ratios drop significantly since there is no filter applied in the decimation process. Out of every 5 data samples, 4 are thrown away, and only 1 is kept. In the third column, data samples are not thrown away, and the 5 data samples are averaged. It can be seen that the compression ratio

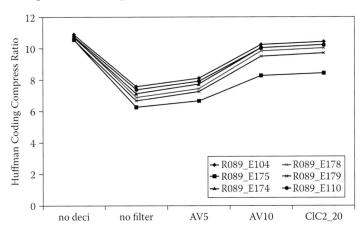

Figure 5.8 The Huffman coding compression ratio.

improve slightly. In the fourth column, the average of 10 data samples is used as the decimation filter. The compression ratio improves significantly since the filter starts satisfying the sampling theorem, and the alias noise is significantly reduced. In the rightmost column, a CIC-2 filter that satisfies the sampling theorem is used as the decimation filter. It shows an even better compression ratio since the stop band suppression of the CIC-2 filter is better than the CIC-1 (the average) filter.

It can be seen that strictly following the sampling theorem is important in noise-sensitive applications such as Huffman coding. This is why we have emphasized this early in this chapter.

References

1. A. Oppenheim et al., *Signals and Systems*, 2nd ed., Upper Saddle River, NJ: Prentice-Hall, 1997.
2. R. Lyons, *Understanding Digital Signal Processing*, 2nd ed., Upper Saddle River, NJ: Prentice-Hall, 2004.

chapter six

Examples of FPGA in front-end electronics

6.1 TDC in an FPGA based on multiple-phase clocks

A broad range of time measurement functions in high-energy/nuclear physics experiments can be implemented in an FPGA directly. There are two types of practical TDC structures in an FPGA with different timing resolutions and complexities. The first TDC scheme is based on a multi-phase clock, and typically the input of the TDC is sampled by four registers with four phases of the clock as shown in Figure 6.1.

In this design, the input is buffered with a logic element, and then sent to four registers with equal propagation delays. The four registers are connected to four internal clocks, each with a 90° phase difference. The 0° and 90° clocks are generated in a phase-lock-loop (PLL) clock synthesizer, and their inversions are used for 180- and 270-degree clocks. If four phases of 500 MHz clocks are used, the input signal is sampled every 0.5 ns, which forms a TDC with 0.5 ns bin size (0.15 ns RMS). Note that the sampling interval is 0.5 ns, but each register operates at 500 MHz, rather than at 2 GHz. A transfer to the 0-degree clock domain occurs in the second and third stages of the pipeline. Depending on arrival time, the transitions of the input logic levels are recorded at different locations within the four registers. The position of the input signal edge being sampled represents the arrival time and is encoded as lower two bits, T0 and T1, of the time value plus a data valid signal, DV. The higher bits TS are generated with a coarse time counter. The coarse time, fine time, and data valid signal are sent to later stages for further zero suppression, buffering, and packing operations.

Transition-edge regulation and detection logics are included in the encoder. For many applications, a simple leading-edge encoding is sufficient. In some applications, for example, to estimate an input pulse charge from a wire chamber, both leading and trailing edges may be digitized. An additional output indicating the type of edge may be needed in this case.

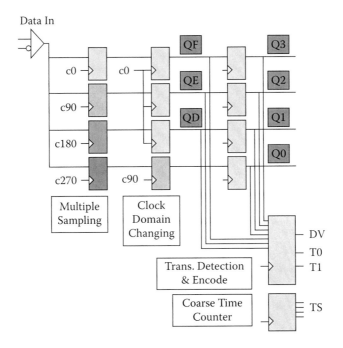

Figure 6.1 Multiple sampling TDC structure in an FPGA.

A function of transition edge regulation prevents ultrashort pulses due to input circuit ringing from being mistakenly digitized. In this design, up to four consecutive bits in the bit pattern QD to Q3 are used by a lookup table in the FPGA logic element to determine if a sampling point is at the edge of a well-established pulse. For example, due to impedance mismatch caused by cable aging, signal reflection in a long cable may produce a bit pattern "0000101" on QD to Q3 with several transitions instead of an ideal pattern, "0000111". One may design edge-detecting logic functions such as Q1&(!Q0)&(!QF)&(!QE) to recognize a subpattern "0001" as a valid transition edge, instead of Q1&(!Q0), which detects any "01" subpatterns as a transition edge. This way, only one valid transition edge will be detected in the bit pattern, even with the presence of input signal ringing. Recall that by using a lookup table in FPGA, one can implement "any" four-input combinational logic, satisfying the edge-detection and pulse-filtering requirements of an application.

Timing critical signal paths are controlled by placing the input buffer, multisampling registers, and clock domain transfer registers in the FPGA to locations that will assure equal propagation delays from input buffer to the sampling registers, resulting in uniform bin widths and thus minimizing differential nonlinearity. An example of placement for the timing of critical logic elements is shown in Figure 6.2.

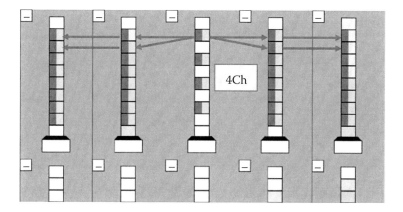

Figure 6.2 Logic elements placement for the multiple sampling TDC.

In each channel, the input buffer cell is placed in the logic array block in the middle, and the registers driven by the clock with four phases are placed in the blocks left and right of the center one to ensure equal propagation delays. Further left and right blocks contain registers for the clock domain transfer. Placement of other logic elements is relatively flexible and can be automatically placed with the compiler.

Placing logic elements "manually" is a time-consuming task, but it is possible to use a spreadsheet to do the work efficiently. The locations of the timing-critical input buffer and flip-flops (about 10 items per channel) for all TDC channels can be kept in the spreadsheet with the FPGA internal coordinates. The designer may further arrange the location of each channel or channel group to adjust the input delay from input pins so that the skews between different channels are minimized. The spreadsheet is coded to output an ASCII file that is pasted into the assignment file for compilation with the FPGA design software.

The obvious advantage of this structure is low resource usage and relatively low sensitivity on temperature and power supply voltage, but the timing resolution is limited by the maximum clock frequency inside the FPGA.

In some early work, four sets of sample, edge detect, pulse filter, and count latch are driven by four clocks with 90° phase separations. These four sets of data collected by four sets of circuits are excessive, and they become valid at different times, which makes the metastability elimination and encoding logic complicated. In the TDC design given above, the four samples are transferred into a bit pattern in a single clock domain immediately, and only one set of edge detect, pulse filter, and count latch circuit is used. The metastability is limited at the sampling stage only, and in fact, the meta-stability in the sampling stage does no harm; it only

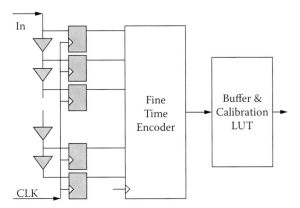

Figure 6.3 Delay chain-based TDC in FPGA.

carries the input signal arrival time information. The decoding becomes very simple in this design as described earlier.

6.2   TDC in an FPGA based on delay chains

Chain structures existing in most of today's FPGA families can be used for TDC purposes (see [1–9]). The structure shown in Figure 6.3 uses a carry chain, followed by a register array in FPGA devices.

The relative timing between the input signal and the internal clock is measured with the position of the transition edge registered in the register array. The structure is commonly used in application-specific integrated circuit (ASIC) TDC chips, except in ASIC chips the delay chain is adjusted by a control voltage that is derived by a feedback loop, so that the delay of each tap is a known constant. In an FPGA, the delay of the delay chain is not controlled, and it changes as the temperature and power supply voltage vary.

Another different between ASIC TDC and the FPGA TDC is that in ASIC devices, the designers can choose to delay either hit input, or clock or both, while usually in an FPGA only hit input can be delayed, and the elements of the register array are driven by a common clock.

A special feature of the FPGA TDC is its large differential nonlinearity (DNL) as shown in Figure 6.4, which is represented as the apparent width of each TDC bin. There are several origins of DNL. (1) The first and the most significant one is the logic array block (LAB) structure. When the input signal in the carry chain passes across the LAB boundaries (and also the half-LAB boundaries in some FPGA families), extra delays added cause periodic "ultrawide bins." Based on measurement, in an Altera Cyclone II device (EP2C8T144C6), the typical raw bin width is about 60ps, while the ultrawide bins can be as large as 165ps. (2) Another origin of the DNL is the delays in the clock distribution network. The clock

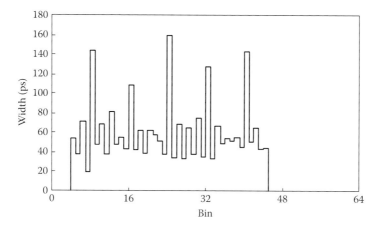

Figure 6.4 Bin widths of delay chain-based TDC in an FPGA.

signals drive different flip-flops in the register array not to be exactly simultaneous. (3) There is also a logic or firmware origin of the DNL. The carry chain in an FPGA is actually a small lookup table allowing users to specify a different carry logic. The delay cells can be specified as either noninverting or inverting buffers. With inverting delay cells, the input signal passes through the delay chain with alternating opposite logic transitions that have different propagation delays causing different widths of the even and the odd bins. In some cases, the DNL of the even–odd bins can be a good feature that will help us to improve the overall measurement resolution.

Two major issues must be solved for the practical FPGA TDC "turn-key" applications. (1) The bin widths are uneven and depend on temperature and power supply voltage, which must be calibrated as frequently as possible. The autocalibration functional block described later provides semicontinuous calibration that converts the TDC measurements from bins to picoseconds. (2) In many applications, the maximum bin width of the ultrawide bins limits the TDC resolution. The "wave union launchers" described later are designed to make multiple measurements with a single delay chain structure, effectively subdividing the ultrawide bins in each raw measurement.

6.2.1   Delay chains in an FPGA

In this section, several considerations of delay chains are discussed. The designers normally cannot redesign the delay chains in an FPGA, so choosing an appropriate FPGA family with a suitable delay chain structure is a crucial step toward a successful TDC design.

The routing between arbitrary logic elements may need to pass several interconnecting matrices and the propagation delays can be very long with large variations even after a laborious hand layout of the logic elements.

Carry chain structures are available in most FPGA families designed for implementing adders, accumulators, and counters for digital processing applications. The carry chains are dedicated routes between FPGA logic elements with minimal propagation delay so that counters with many bits can operate at high frequencies. Therefore, it is recommended to use a dedicated carry chain structure for the TDC instead of using generic interconnects between logic elements.

It should be pointed out that a carry chain that is too fast is not suitable for TDC implementation purposes. An ideal delay chain should have relatively uniform propagation delays in each delay cell so that the differences of the input signal arrival times can be recorded in the register array.

In some high-end FPGA families, advanced carry generation schemes such as carry selection may be utilized to optimize the performance of high-speed adders. Implementing TDCs using these families is significantly difficult, if not impossible, compared to implementing TDCs using low-cost families with plain carry chains.

The delay line length should be kept as short as possible to reduce both logic resource usage in the encoder and also the measurement errors, especially at the middle of the delay chain. To reduce delay chain length, the clock frequency driving the register array should be chosen as high as can be reasonably achieved. Typically, at a relatively high frequency in an FPGA, the delay chain length is around 32 to 64.

Different logic resources in FPGAs usually have different maximum operating frequencies. The high frequency chosen for the register array is likely to be too high for circuits in later stages, especially memory blocks. Schemes interfacing a fast register array with a slow back stage are normally necessary.

6.2.2   Automatic calibration

It is known that the propagation delay of a delay cell depends on temperature and power supply voltage. In ASIC TDC it is possible to compensate the delay variation using the analog method, that is, to generate a control voltage from the phase difference of external crystal oscillator and the internal ring oscillator, and to use the control voltage to fine-tune the internal cell delays via a negative feedback. In an FPGA, instead of making compensation, the propagation delay of the delay chain is measured in real time. Using this delay value, the actual arrival time of the signal can be found either offline or online inside the FPGA, using a lookup table.

In an FPGA TDC, analog compensation is not convenient and digital calibration is more preferable. There are at least two approaches to digital calibration—the average delay approach and the bin-by-bin approach.

In the average delay scheme, the total delay time of the delay line is designed to be longer than the clock period t_p. Sometimes, an input logic transition will be recorded by the register array twice. If the positions of the two registered logic transitions are N_1 and N_2, respectively, then the average cell delay is:

$$t_d = \frac{t_p}{N_2 - N_1} \tag{6.1}$$

Sometimes, the number of delay taps propagated in a clock period can be viewed as a fractional value rather than an integer. This value is calculated from multiple measurements, and will provide a more accurate calibration.

The advantage of this scheme is its fast response time. However, it does not provide bin-by-bin calibration when the bin widths are different, since only the average cell delay is measured in this scheme.

For FPGA-based TDCs, bin-by-bin calibration is recommended since the widths of the bins vary by a large range.

Assuming that the widths of all TDC bins are measured and stored in an array w_k, then the calibrated time t_n corresponding to the center of bin n can be written as

$$t_n = \frac{w_n}{2} + \sum_{k=0}^{n-1} w_k \tag{6.2}$$

It should be emphasized that it is crucial to calibrate to the centers of the bins; that is, the first term representing the half width must not be omitted. It is not impossible for one to implement the sum term only and omit the half-width term when the calibration algorithm is buried in complicate codes.

It can be shown that the RMS measurement errors are the minimum when the times are calibrated to the centers of the bins. Consider the RMS error σ contributed by one bin with lower and upper limits of t_1 and t_2, respectively. If this bin is calibrated to a value t_c between the lower and upper limits, the contribution of the error can be written:

$$\sigma^2 = \frac{1}{(t_2 - t_1)} \int_{t_1}^{t_2} (t - t_c)^2 \, dt = \frac{(t_2 - t_c)^3 - (t_1 - t_c)^3}{3(t_2 - t_1)} \tag{6.3}$$

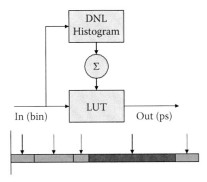

Figure 6.5 The automatic calibration block.

When the bin is calibrated to the center, that is, $t_c = (t_1 + t_2)/2$, the error above reaches a minimum which is $(t_2 - t_1)^2/12$.

The sum term in Equation 6.2 represents the calibration to the edges of the bins. When all the bins have identical widths, the half-width term is a constant offset, and calibrating to either bin edges or to bin centers will result in the same RMS errors. However, when the widths of the bins are different, the RMS errors will increase. The automatic calibration functional block for bin-by-bin calibration is shown in Figure 6.5.

After power-up or system reset, all TDC inputs are fed with calibration hits. The timing of these hits should have no correlation with the clock signal driving the TDC, so the hits should be generated from an independent oscillator. It is also possible to use real event hits as calibration hits if the hit rate of the real events is sufficiently high.

The input from the TDC encoder for a 64-tap delay line/register array structure is a 6-bit number, representing the bin number of the logic transition of the input signal with a possible range of 0 to 63. A 64-bin DNL histogram is booked in the FPGA internal memory. If the number of total hits is known, then the counts in each bin can be used as its bin width. For example, if 16384 hits are booked into the histogram and assume these hits are evenly spread over 2500ps, the period of the 400MHz clock driving the TDC, then the width of a bin with N count is $N*2500ps/16384 = N*0.1526ps$.

Once all hits are booked into the histogram, a sequence controller starts to build the lookup table (LUT) in the FPGA internal memory. The LUT is integrated from the DNL histogram so that it outputs the actual time of the center of the addressed bin. The process is as follows:

1. Half of the width of the first bin becomes the time at its center.
2. Another half-bin width of the first bin and the half-bin width of the second bin are added to get the center time of the second bin.
3. This sequence is repeated for remaining bins.

Once the LUT is built, the outputs of the LUT are the TDC times calibrated to the temperature and power supply condition during booking the previous set of hits.

In normal operation, a new DNL histogram can be booked as real event data are taken. Each time a new DNL histogram is booked with 16K hits, a new calibration LUT can be built and used for subsequent events. In real implementation, the DNL histogram booking/LUT building process and the current LUT are usually in the same physical memory block with different memory pages.

6.2.3 The wave union TDC

The automatic calibration described above provides times corresponding to the center of each bin, but it will not change bin size. The issues of the ultrawide bins due to uneven physical structure of the carry chain inside an FPGA are still to be addressed. The wave union TDC, consisting of a wave union launcher feeding a delay chain/register array structure, is developed to subdivide the ultrawide bins. A wave union launcher creates a pulse train or "wave union" with several 0-to-1 or 1-to-0 logic transitions for each input hit and feeds the wave union into the TDC delay chain/register structure, making multiple measurements. An example of the wave union launcher implemented in a logic array block with 16 logic elements is shown in Figure 6.6. It is connected with the remaining 48 cells in the 64-cell carry chain/register array.

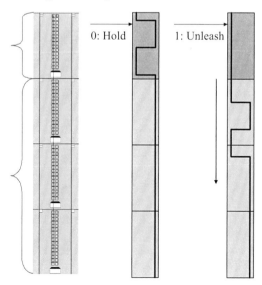

Figure 6.6 The wave union TDC.

When the input level is 0, a logic pattern or "wave union" with two 1-0 transitions and a 0-1 transition is formed in the launcher, and the pattern is held in place. When the input level becomes 1, the wave union is unleashed to propagate in the carry chain. At the leading edge of the clock, a snapshot is recorded in the register array, and the position of the wave union represents the arrival time of the input signal.

The nominal separation of the two logic transitions in this example is 13 bins, and they are encoded for further processing. The period of the ultrawide bins of this FPGA device is 8, and at least one of the two logic transitions will be in a normal bin. This arrangement effectively subdivides the ultrawide bins or improves the sensitivity of the TDC. If one transition is in an ultrawide bin and is not sensitive to the arrival time change, the other transition will be in normal bins and maintain the sensitivity. The effect can be seen in Figure 6.7.

One of the two logic transitions is encoded, and its DNL histogram is booked and marked with "Plain TDC" in which the ultrawide bins are seen. The sum of the bin numbers of the two transitions is also booked and marked with "Wave Union TDC A." Now as expected, no bin is ultrawide anymore. The sum of the bin numbers spreads from about 20 to about 100, about twice of the range for plain TDC.

When both of the transitions are in normal bins, nominally they are in opposite odd–even bins, given their separation of 13. When the odd and even bins have different widths, they effectively subdivide each other and, therefore, further improve the TDC resolution. However, the resolution improvement primarily comes from eliminating the ultrawide bins.

It is interesting to compare this scheme with the interleaving scheme in ASIC TDC for the improvement of resolution. Both schemes use multiple measurements to reduce measurement errors. But in this design, the

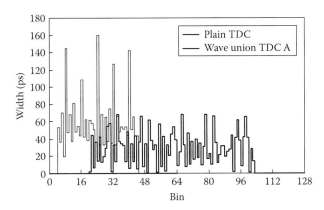

Figure 6.7 The bin widths of wave union TDC and plain TDC.

two measurements are made in the same delay line/register array structure to save logic resources. In ASIC, since fine delay control is possible, usually the time difference of the two TDC inputs is exactly half of the bin width, which yields a best gain of factor of 2 for resolution improvement. In an FPGA, fine propagation delay control is very hard, and the resolution improvement is more or less a factor of 1.4 (square root 2), if there are sufficient variations of the bin widths. This is the reason why it is preferable to use inverting buffers in the carry chain, which causes an even–odd bin width difference. In this design, improving resolution is a secondary purpose, with subdividing the ultrawide bin being the primary one.

6.3   Common timing reference distribution

The TDC discussed before measures relative timing difference between the input signal and a common internal clock signal. In a large system when multiple TDC modules are needed, there are two approaches to synchronize the system. Traditionally, the clock signal is distributed in the system, and on each module the clock is regenerated using phase-lock-loop (PLL) circuits. This is an approach to distribute the timing reference using the analog compensation method. In a pure digital approach for TDC systems, a pulse or a set of pulses with common timing transitions are distributed, and the arrival times of the transitions are measured with TDC channels dedicated for common timing purposes. The arrival times of the common timing pulses represent relative timing offsets between different TDC modules, and the offset is simply subtracted for system synchronization.

6.3.1   Common start/stop signals and common burst

Traditionally, TDC devices measure the time difference between the input signals and a common timing reference signal. Depending on whether the common timing signal is earlier or later than the inputs, the timing signal is referred to as "common start" or "common stop."

In an FPGA TDC, the timing reference channel consists of a regular TDC and a functional block that sums up the times from a pulse burst. The TDC digitizes the rising edges of the input pulses, and the times of the pulses are fed into the burst sum block. The burst sum block can be designed to accept inputs of the timing reference channels with one or more pulses.

In the special case when the burst is just a single pulse, the common timing reference is similar to the traditional common start scheme. In this case, the arrival time of the common timing reference is reported in the data stream and the time intervals between the common start signal and the individual channels hits can be calculated with a subtraction. Note that neither the common "start" pulse here needs to arrive earlier than the channel hits, nor are the channels "stopped" after being hit.

The primary motivation of implementing the burst sum function in the timing reference channels is to support advance timing distribution schemes. In conventional common start/common stop schemes, the common timing signal is distributed in a single shot, suffering circuit jitter and binning errors in TDC. In this design, the reference time inside an FPGA is an average of multiple (up to 8) measurements, and multiple measurements provide finer timing resolution than a single shot.

In the common burst mode, a burst of 2, 4, or 8 pulses are used as a common timing reference signal. The average of times of pulse rising edges is reported to the data stream. With an average of four measurements, for example, timing jitter is reduced by a factor of 2 and an additional bit of the timing resolution is anticipated.

6.3.2   The mean timing scheme of common time reference

A very attractive timing distribution method is the mean timing scheme; the mean timing scheme is a special case of common burst mode.

The timing distribution system drives a multidrop copper twist pair cable from both ends as shown in Figure 6.8a. The left and right end drivers are alternatively enabled and drive pulses to travel from left or right end. There is no need to synchronize the pulses. The pulses from left and right drivers can be at any arbitrary times. The differential signals are received in each TDC module/FPGA, and the arrival times are digitized.

The mean timing burst has 8 pulses as shown in Figure 6.8b. The receivers on each TDC FPGA receive the burst with 4 pulses delayed from the left path and 4 from the right path. The traces represent pulses seen at different TDC modules. The arrival times at different modules are different, but the mean times of the 8 pulses as indicated with the dots are the same.

The only required condition in this scheme is that the cable segments have the same propagation delays for left-going and right-going pulses. There is no requirement on actual values of the delays and temperature variations and, therefore, no requirement of using high quality media. Any moderate-grade media like Cat-5 twist pair cables or even ribbon cables can serve this purpose. The TDC firmware that supports the common burst mode can also support the mean time mode without any changes.

6.4   ADC implemented with an FPGA

Intrinsically, the FPGA is a digital device. However, with the suitable use of FPGA resources, it is possible to use the FPGA to digitize analog waveforms. The digitized waveforms can be directly processed in the FPGA.

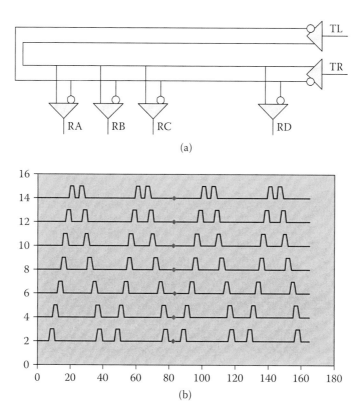

(a)

(b)

Figure 6.8 The mean timing scheme: (a) the left and right end drivers and the receivers, (b) pulses at different receivers and the mean times.

There are several possible schemes of digitizing analog signals suitable for FPGA implementation with different advantages and drawbacks, which will be discussed in the following text.

6.4.1   The single slope ADC

A simple scheme of ADC implemented in FPGA for digitizing multiple channels is based on analog-to-time conversion, comparing a ramping reference voltage with the input waveform as shown in Figure 6.9.

In this scheme, the analog inputs are directly connected to the FPGA input pins. A passive RC network is connected to the FPGA output pins so that a periodic reference voltage ramp can be generated. The differential input buffers in an FPGA are used as comparators to generate logic transitions inside the FPGA when the reference voltage ramps across the input voltage levels. The transition times are digitized using the TDC block implemented in the FPGA. Since the period, the RC network parameters,

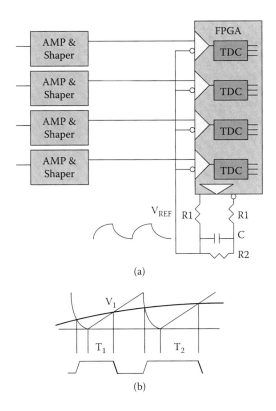

(a)

(b)

Figure 6.9 The single slope ADC implemented with an FPGA: (a) block diagram, (b) an example of the ramping reference voltage.

and the starting time of the ramps are known, the input voltage levels can be derived from the transition times.

In today's FPGA devices, differential input buffers are good comparators within a sufficiently large range of input voltage levels, since they are designed to be compatible with various differential signaling standards. The ramping-comparing scheme we studied here is a suitable choice for applications with a large channel count of relatively slow signals. This scheme is classified as a single-slope ADC, although both ramping slopes can be utilized. In some references, the single-slope scheme is mistakenly referred to as a Wilkinson ADC that is based on the dual-slope principle.

The ramping reference voltage can be generated with other circuits to have different ramping shapes for optimal performance of the ADC, and an example is shown in Figure 6.9b with the up-going ramp being approximately linear. Note that, however, there is no requirement to design the ramping reference voltage as linear. Inside the FPGA, output times from the TDC can be easily calibrated to actual voltage through a

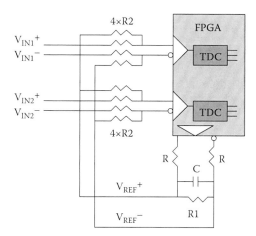

Figure 6.10 The single slope ADC with differential inputs and reference voltage.

lookup table, taking the nonlinear shape of the ramping reference voltage into account.

The single slope ADC can also be designed to accept differential analog inputs, and the reference voltage can also be differential as shown in Figure 6.10. In this circuit, the resistor network for each input sums up the input and the reference voltages. The ADC with differential inputs rejects a common mode in inputs and is suitable for applications sensitive to external noises.

6.4.2   The sigma-delta ADC

The sigma-delta ADC uses a comparator and an integrator to generate 1-bit digitization data that oversamples the input. The high and low times of the comparator are recorded and are used to calculate the input voltage level. A very simple version of the sigma-delta ADC suitable for FPGA implementation is shown in Figure 6.11.

A differential input pin pair is used as a comparator that compares the input voltage and the reference voltage generated from an FPGA output

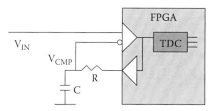

Figure 6.11 The sigma-delta ADC implemented with an FPGA.

pin with an RC network. The RC time constant is chosen to be relatively small so that the reference voltage can track the input voltage relatively fast. Times of both the rising and falling edges of the comparator are digitized in the TDC. With sufficient TDC data, the input voltage level can be calculated fairly accurately.

Differential versions of the sigma-delta ADC can also be designed using a circuit similar to Figure 6.10 for noise-sensitive applications.

The sigma-delta scheme given here uses more FPGA pins and external components than the single-slope ADC scheme, and the calculations for converting TDC data to voltage level are more complicated. The advantage of the sigma-delta scheme is that it has a relatively fast response to the input voltage change, although at a coarser precision. On the other hand, higher precision for a slow-changing input signal can be achieved with more measurements over a longer time. So the sigma-delta ADC is suitable for applications that demand trading off the measurement speed and accuracy in the field dynamically, while the single-slope ADC is more suitable for applications with fixed speed and accuracy.

6.5   DAC implemented with an FPGA

If analog voltages are to be produced from the digital data inside the FPGA, a digit-to-analog conversion (DAC) is needed. Typically, a separate chip is used to perform the DAC function, and the DAC chip is connected to the FPGA via either parallel or serial digital interface. In some applications, if only a few channels of slow-changing analog voltage are needed with moderate precision, it is convenient to simply implement the DAC directly with the FPGA. For each DAC channel, only one FPGA pin is needed, plus a few external resistors and capacitors functioning as a low-pass filter. In this section, two approaches to DAC implementation are discussed.

6.5.1   Pulse width approach

The pulse-width-based DAC consists of a counter and a comparator inside the FPGA as shown in Figure 6.12a.

Consider an example of a DAC with an 8-bit counter and an 8-bit comparator. The counter repeatedly counts from 0 to 255, and its value is compared with the DAC input. If the DAC input is an integer N, then the output of the comparator is high when the counter output is from 0 to $N-1$ so that the width of the pulse is N clock cycles as shown in Figure 6.12b. Note that if $N=0$, there is no pulse, that is, the pulse width = 0. After being filtered by the RC network, the output becomes a smoothed voltage level that is approximately proportional to the duty cycle of the output pulse. If the bank of the output pin of the FPGA is powered to 2.5 V, the output range of the DAC can range from 0 to approximate 2.5 V.

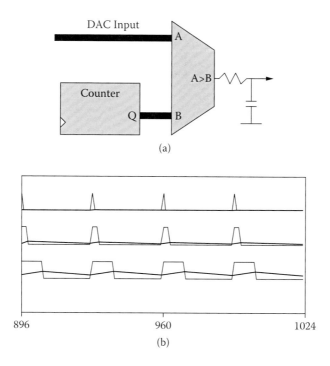

Figure 6.12 Pulse-width-based DAC implementation: (a) block diagram, (b) outputs with different pulse widths and voltage levels after the RC filter.

If the clock frequency driving the counter is f, the output of the DAC is a periodic signal with primary AC noise concentrated at frequency $f/256$. This DAC scheme is suitable for using a notch filter to suppress the fundamental frequency peak along with a regular low-pass filter.

In fact, the LED brightness variation scheme described earlier is based on this DAC scheme in which the low-pass filtering is performed by human eyes.

6.5.2   Pulse density approach

Another approach to DAC implementation is based on pulse density, using an accumulator with a carry output as shown in Figure 6.13a. In this scheme, pulses of one clock cycle wide are approximately evenly spread. The number of pulses in a given time period is proportional to the DAC input. The total times of high and low outputs in this scheme are the same as in the pulse width approach, except that the high-level outputs are distributed as single clock period pulses rather than combined into a wide pulse. After a low-pass filter, the output voltage is approximately proportional to the DAC input.

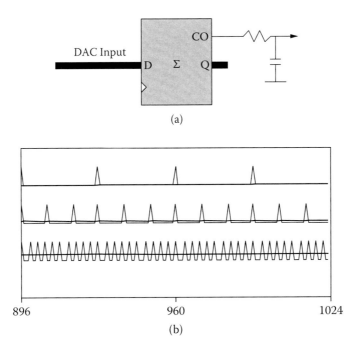

Figure 6.13 Pulse density-based DAC implementation: (a) block diagram, (b) outputs with different pulse densities and voltage levels after the RC filter.

Again, consider an 8-bit DAC for simplicity, and assume the DAC input is N. Should the accumulator have 16 bits, then after 256 clock cycles, it would become $N \times 256$. For an 8-bit accumulator, it will provide a carry bit that becomes high for a total of N clock cycles, and these N clock cycles are approximately evenly distributed as shown in Figure 6.13b.

Let us use an actual number, $N = 10$, for a better explanation. If at the starting point, the accumulator register is cleared, that is, $Q = 0$, then after 25 clock cycles, $Q = 250$. During the 26th clock cycle, carry out CO $= 1$, and Q becomes 260, but its lower 8 bits roll over to 4. After another 26 clock cycles, another carry-out pulse is generated, and Q becomes 264, which rolls over to 8. Next, after 25 clock cycles, the third carry-out pulse is generated while Q becomes 258, which rolls over to 2. So when $N = 10$, a carry-out pulse is generated every 26 or 25 clock cycles and after 256 clock cycles, 10 carry-out pulses are generated.

The lowest AC noise frequency is at $f/256$, which is the same as in the pulse width approach. However, if N is an even number, this $f/256$ peak vanishes and the lowest AC noise frequency becomes $2(f/256)$, which vanishes if N is an integer multiple of 4. For other N values, the noise spectra also tends to concentrate to higher harmonics. So this DAC approach is suitable for applications with simple low-pass filters.

For both DAC approaches, the sampling rate of the DAC can be defined as $(f/2^B)$ where B is number of bits of the DAC. If the clock frequency is 200 MHz, a 12-bit DAC will have a sampling rate of (200 MHz)/(4096) = 48.8 kHz, which is sufficiently high to generate voice signals.

6.6   Zero-suppression and time stamp assignment

Sending raw TDC or ADC data out of an FPGA would normally require too much bandwidth. In fact, the ADC or TDC channel is not hit every clock cycle. Therefore, it is possible to suppress the clock slot that contains no hit data.

Consider a front-end digitizer without a trigger as shown in Figure 6.14, in which all hits are sent to later stages. In the zero-suppression process, a time stamp (TS) must be attached to the hit data to identify which clock cycle the hit data is generating. In the case of TDC, the TS bits are those of a coarse-time counter. The width of the TS is always a debate in nearly every experiment. A time stamp of k bits can represent up to 2^k clock cycles. If the TS is too short, the counter rolls over, resulting in ambiguity about the hit time in integer multiple of 2^k clock cycles. This is a similar problem as the Y2K bug, so we will call the time period of 2^k clock cycles a "centenary."

Increasing k is a possibility, but it costs data link bandwidth. For example, a 32-bit time stamp can represent a time period of 85 seconds with a 50 MHz clock. However, every hit must be attached with a 32-bit number, while the hit data itself may be just a few bits.

Another possibility is to use shorter TS and send a "centenary mark" (CM) when the counter rolls over. For example, an 8-bit time stamp can be used with the TS counter counting from 0 to 254. The value 255 is reserved as the centenary mark. When the TS counter reaches 254, the FPGA inserts a fake hit data in the data stream with time stamp value 255 if there is no real hit at this clock cycle. (If there is a real hit at TS = 254, the time stamp

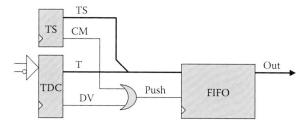

Figure 6.14 Zero-suppression for nontrigger front end.

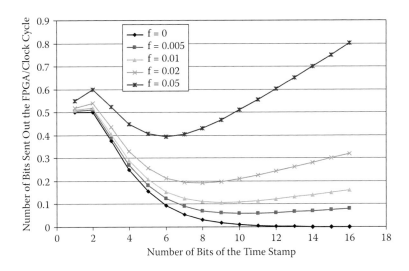

Figure 6.15 Data output rate from the FPGA.

value is 254 to indicate that it is a real hit data.) The receiving devices use the centenary marks to increment the upper bits of the TS counter.

In this scheme, the total number of bits sent out to the FPGA in T clock cycles can be written as

$$N = \frac{T}{2^k} k + Tfk$$

(6.4)

The parameter f is the hit rate, defined as number of hits per clock cycle. The results are plotted in Figure 6.15 for hit rates $f = 0.005$, 0.01, 0.02, and 0.05 hits/clock cycle.

It can be seen that with a hit rate of around 1%, the FPGA output rate is minimum when the number of bits used for the time stamp is 8–12. A long time stamp is only reasonable when the hit rate is extremely low.

The choice of the time stamp also depends on other factors like the time needed for the accelerator turn. For example, an accelerator turn in Fermilab Tevtron is 159 clock cycles at clock period 132 ns. It is more convenient to choose mod 3, mod 53, or mod 159 counters for corresponding bits of the TS counters.

6.7   *Pipeline versus FIFO*

Pipeline and FIFO buffers are two popular types of memory organization methods utilized in high-energy physics triggers and DAQ systems, although the names may not reflect the actual properties of the two buffer

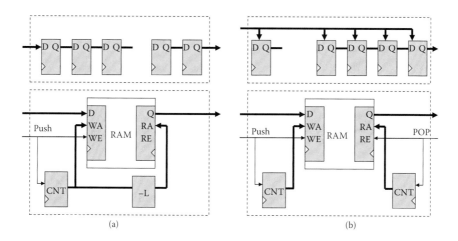

Figure 6.16 Buffers: (a) Pipeline; (b) FIFO.

types. In fact, the data stored in and retrieved out of a pipeline is also in the first-in-first-out category. In this document, we use "pipeline" to refer a buffer with constant storage to retrieve steps, which can be visualized as a serial-in-serial-out shift register. A FIFO buffer is a storage feature where data can be pushed into it and popped out with the same data orders for push-and-pop operations. The pipeline and FIFO buffers are shown in Figure 6.16.

In a FIFO buffer, the write address (WA) and read address (RA) are kept with two counters. The write-enabled (WE) signal is derived from the PUSH signal, which also increases the WA counter. In the output side, the read-enabled (RE) signal is derived from the POP signal that also increases the RA counter. Some varieties of the FIFO may have logic to check for empty (i.e., WA = RA) or full (i.e., WA-RA = number of RAM words –1) conditions. Sometime, additional logics are added to prevent the outside circuit from pushing into a full, or popping out an empty, FIFO.

The pipeline buffers can be viewed as shift registers, but actually they are rarely implemented with shift registers chained up with flip-flops. The flip-flops are not efficient to store data not only in FPGA, but also in ASIC chips. Implementing long pipeline buffers with shift registers unnecessarily consumes a large amount of silicon resources. Also, when data are clocked through flip-flops, transistors are turned on and off in all steps, causing a large power consumption.

The actual implementation of a pipeline buffer is implemented similarly as FIFO, except that the read address RA is derived from the write address WA, with WA – RA = L, where L is the length of the pipeline. The RAM cell uses a lot fewer transistors than the flip-flop. After data

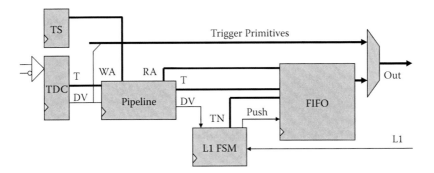

Figure 6.17 A possible trigger front-end.

is stored in RAM, the transistors of the storage cell keep the on or off states unchanged until they are overwritten next time, instead of changing every clock cycle as in flip-flops; therefore, the power consumption for RAM is significantly lower.

In high-energy physics trigger and DAQ systems, pipelines are often used to store detector data for a fixed number of clock cycles waiting for an L1 trigger. The FIFO buffers, on the other hand, are used when the instantaneous data rate is considerably different from the average data rate, such as in the zero-suppression process. A possible triggered front-end design that uses both pipeline and FIFO buffers is shown in Figure 6.17.

In this model front-end FPGA, the arrival times of the detector hits digitized in the TDC block with the DV signal, indicating that a valid hit is detected at the given clock cycle. The bits of T representing the arrival time of the detector hit along with the DV signal are written into the pipeline buffer every clock cycle (regardless, if there is a hit or not) and the time stamp TS is used as the write address, WA. The immediate hit information—perhaps, just the DV signals of all channels—is sent into the output data stream as trigger primitives.

Of course, the trigger primitives can also be more complicated. For example, the mean time of the arrival times of two adjacent channels can be used, which represents the particle track hit time or event time (plus a constant delay) in drift chamber cases.

The trigger primitives are collected by the trigger/DAQ system and are used to generate a global level 1 trigger L1.

The L1 returns back to the front-end at the time when the hits from the trigger event are about to reach the end of the pipeline. Conducted by the L1 finite state-machine (FSM), nonempty data are pushed into the FIFO buffer. Now, a DV signal is used to derive the PUSH signal so that only nonempty data are pushed. However, additional bits must be included in the hit data. Usually, a trigger number (TN) or something similar is put into the trigger packet data header. For every hit, several bits of the read address RA that represent the coarse time must be added. The hit data pushed into the FIFO are sent out to the trigger/DAQ system when the output channel is not used to send the trigger primitives.

Also, it is not necessary to require the L1 return latency to be a constant. The L1 trigger can be a multibit command, and several bits in the command can be assigned as the starting time stamp of the L1 window. The L1 FSM generates the corresponding read address RA that will ensure the hit data in correct timing window are collected. This scheme can be used in trigger systems when variable L1 trigger latency is necessary.

We shift our attention back to the comparison of the pipeline and the FIFO buffers. In the pipeline shown in the previous example, many time slots contain no valid hits. It seems that zero suppression should be done right after the TDC, rather than after receiving the L1 trigger. In other words, it appears to be more economical to replace the pipeline in Figure 6.17 with FIFO.

However, several factors must be considered while choosing a zero-suppression stage. First, the zero-suppression process increases data word width, since the time stamp must be added, while in the pipeline buffer, the read address RA itself is the time stamp. Second, in the FIFO used for zero suppression, a WA and a RA counter must be kept for every channel, whereas in the pipeline buffer the WA and RA are common for all channels.

It is certainly possible to implement the fix latency pipeline with zero suppression using FIFO. A block diagram is shown in Figure 6.18. When the hit data is pushed into the FIFO, the time stamp TS is also stored along with fine time T. At the output side, the TS value of the last hit is compared with the current TS. When the last hit is older than the predefined pipeline length, the POP operation is performed so that the RA in the FIFO points to the next newer hit.

It must be pointed out that the FIFO full is an error source, in addition to other possible error sources in the entire front-end and trigger/DAQ system. In a system with a low hit rate, a sufficiently deep FIFO can reduce the probability of an FIFO full error to nearly zero. However, as

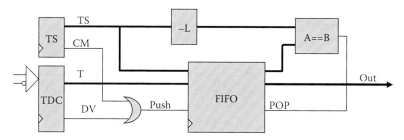

Figure 6.18 Implementation of pipeline using FIFO.

long as the probability is not zero, an error code and all corresponding error handling processes must be implemented. To completely eliminate the possibility of the FIFO full error, the FIFO depth should be bigger than the pipeline length. If so, the FIFO uses more memory space than the plain pipeline without zero suppression, and therefore, there is no advantage in performing zero suppression at all in this situation.

6.8   Clock-command combined carrier coding (C5)

The FPGA TDC and commercial ADC allow designers to place digitization functions in close proximity of the detector. When the digitization is done near the detector, delicate analog or timing signals will not need to be sent over long cables. However, necessary supports must be appropriately planned for the front-end digitization devices.

Obviously, the digitization FPGA must be clocked. At the beginning of a run, there may be some registers or parameters to be set in the FPGA, which requires a means of command transmission. Before data taking, the first clock cycle needs to be marked so that the time stamp counter can be properly started. During normal operation, trigger acceptance commands that start data downloading are to be sent to the front-end for triggered experiments.

The Clock-Command Combined Carrier Coding (C5) scheme was developed to send commands, especially synchronized ones like the first clock cycle marker with a clock signal using a single link. In the C5 scheme, all leading edges of the pulses are separated with an equal distance as a regular clock signal, while the data are encoded into the pulse width. Therefore, there is practically no "recovery" needed for the clock. The pulse train can drive sequential logics directly just as an ordinary clock signal. The C5 pulse trains are DC balanced, suitable for the AC-coupled transmission media.

In the C5 scheme, the channel initiation processes such as preamble, training pattern, frame synchronization pattern, etc., are not needed. This simplifies the design of both the sender and the receiver.

When there is no message sending, the pulse is a plain 50% duty cycle clock. To carry message bits, pulse widths become wider or narrower while all the leading edges of the pulses remain the same clock times.

6.8.1   The C5 pulses and pulse trains

Consider each clock period to be 4 unit intervals (UIs) long. A narrow pulse has 1 UI high and 3 UIs low with 25% duty cycle; a wide pulse has 3 UIs high and 1 UI low with 75% duty cycle, and a normal 50% one has 2 UIs in both high and low times. We use (−1), (0), and (+1) to denote three possible widths of each pulse, that is, narrow, normal, and wide, respectively.

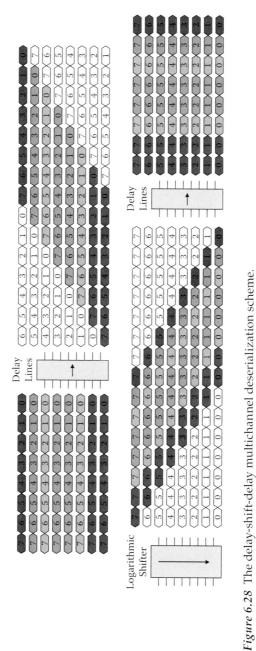

Figure 6.28 The delay-shift-delay multichannel deserialization scheme.

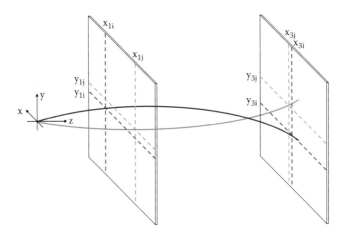

Figure 7.3 A hit pairing problem.

Consider a pulse train with 5 pulses. Given that there are 3 possible widths for each pulse, the total number of possible combinations is 35 = 243. Out of these combinations, 51 of them are DC balanced.

We further require that the first pulse is a nonzero one, that is, either wide (+1) or narrow (−1). This requirement ensures simple frame detection. After a long time with no message transition, the first nonzero pulse represents the start of a message frame. This requirement reduced the number of combinations down to 32.

Up to this point, one may transmit a 5-bit frame in 5 clock cycles. We denote this coding scheme as "5B/5C." However, it is preferable to transmit message frames that fit byte boundaries. In order to transmit 4-bit frames using 5 clock cycles (4B/5C), we select fewer combinations.

Since the first pulse is either a wide or narrow one, a disparity is created. We require that the next nonzero pulse should cancel this disparity. The number of combinations that meet this requirement now becomes 24.

We assign 16 of them as "data codes" as shown in Figure 6.19a to transmit 4-bit frames. The other 8 are called "control codes" as shown in Figure 6.19b, which have the same maximum disparity and therefore are also suitable for transmitting information. However, the coding scheme is so simple that there is practically no control protocol needed in the operation.

6.8.2   The decoder of C5 implemented in an FPGA

The encoder of C5 coding is fairly straightforward, and it can be implemented using a lookup table in an FPGA to create pulses with predefined widths. As in most other coding schemes, the decoder design in the receiving end needs to be developed more carefully to take full advantage of the coding scheme for best performance and lower resource usage. A possible decoder design is shown in Figure 6.20.

The input pulse train CC to the decoder can drive sequential logic and even phase-lock-loop (PLL) circuit directly. A recovered ×4 clock is generated as a system clock for other functions in the chip and used to provide delays for decoding functions. The decoder given here is able to decode 16 data codes and uses only 22 logic elements.

The decoder extracts pulse width information by sampling two delayed versions of the pulse train. The two samples are pipelined down in two shift registers. The value of a bit is derived from the width of a pulse directly using a simple AND (or NAND) operation.

The decoder is self-framed when a transmission starts. Initially, after the decoder receives more than 5 plain clock pulses, the (mod 5) counter is held in the reset state. When the first wide or narrow pulse appears as the first pulse of the 5-pulse train, the counter begins to count. When the counter reaches 4, the pulse train is correctly aligned in the shift register

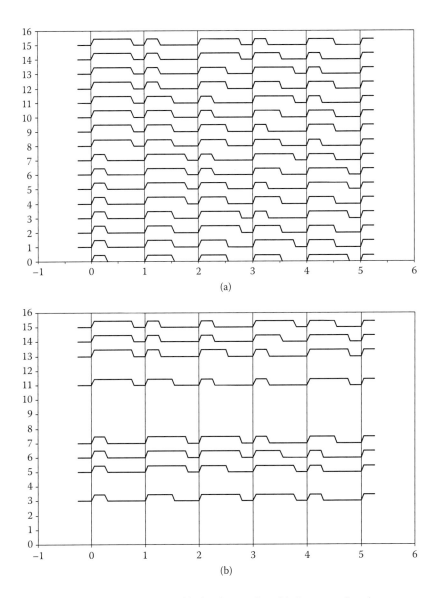

Figure 6.19 The C5 pulse trains: (a) the data codes, (b) the control codes.

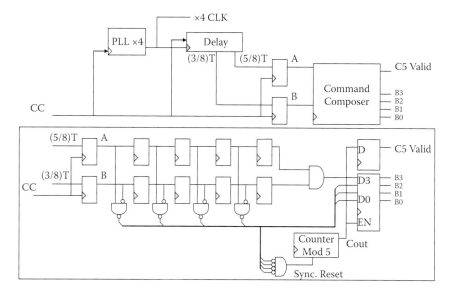

Figure 6.20 The C5 decoder.

and the data is stored in the parallel holding register as a data valid signal "C5Valid" is generated.

The counter rotates to 0 to allow the next data frame to be received immediately. Arbitrary numbers of data frames can be transferred back to back. The availability of a data frame is signified by the "C5Valid" signal. After the last data frame in a transmission, plain clocks are sent. The decoder is held in the reset state again, ready for the next transmission.

The C5 scheme also contains some intrinsic redundancy that can be used as error detection. For example, the pulse train seen at the receiver end must be DC balanced in the 5-pulse frame. It takes only a few logic elements to implement the error detection feature.

6.8.3 Supporting front-end circuit via differential pairs

When cabling is limited, it is economical to combine into the clock channel commands that otherwise need separate links.

It is natural that carrying information in the clock may cause jitters. In fact, the PLL blocks in many FPGA families permit the different duty cycles used in the C5 scheme. Experiments show that there is no visible instability caused by carrying information in the input clock to the FPGA. As long as the potential jitter is lower than that required by the application, it should not be an issue. For example, when a 1 ns bin size TDC is implemented in the FPGA, the RMS contributions to the measurement error for jitters up to 100 ps are practically negligible.

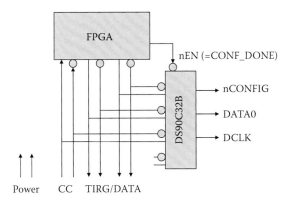

Figure 6.21 A possible scheme of supporting a front-end circuit via differential pairs.

Also, commands such as register setting and parameter loading are typically used in the system initialization stage. The marker of the first clock is only sent once before real data taking occurs. The only possible commands being sent during data taking are triggers, which always happen after the event of interest has been stored in the pipeline (Figure 6.21).

Since the front-end support is discussed, we briefly describe a possible scheme of supporting a front-end circuit via a 4-pair cable such as a Cat-5 twist pair cable with RJ-45 connectors. In this scheme, a pair in cable is reserved to supply power to the front-end circuit board. In a normal operation, a differential pair CC carries clock and commands using a C5 scheme, while the remaining two pairs are used to send trigger or DAQ data out. On power-up, the FPGA to be configured is supported with a LVDS to TTL converter DS90C32B or similar device. Before the FPGA is configured, the converter is enabled and the FPGA pins are tri-stated; that allows signals for configuration to be sent via the differential pairs of CC and TRIG/DATA. After the FPGA is configured, the DS90C32B is disabled, and all the differential pairs resume normal definitions.

6.9   *Parasitic event building*

Event building is a necessary process in all DAQ systems, as well as many trigger systems in high-energy physics experiments. Event building is merging data of the same event from several different subdetectors. When the operation rate of a detector increases, it is often necessary to distribute data of different events to different postprocessors.

In today's trigger/DAQ systems, event building is done primarily utilizing data switches to perform the merging and distributing functions.

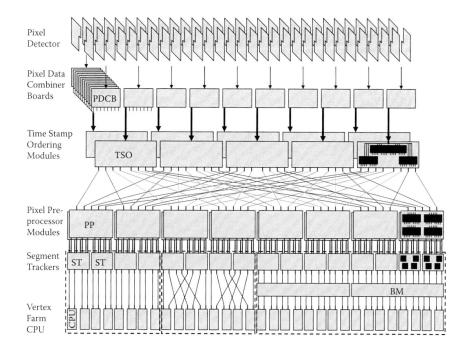

Figure 6.22 Parasitic event building in a proposed trigger system.

However, it is possible to spread the event-building functions parasitically into preprocessing stages or even simple electronics modules like optical receiver or fan-out units. This way, a dedicated data switch can be eliminated. As an example [10], we review a proposed architecture of the Fermilab BTeV level 1 pixel trigger system as shown in Figure 6.22.

We will omit a detailed functional description of various modules and only concentrate on the event-building features. It can be seen that the Time Stamp Ordering (TSO) modules and the Pixel Preprocessor (PP) Modules are cross-connected. Each TSO module has a data path to any of the PP module in the later stage. In normal operation, all events in a TSO are distributed to 8 PP modules, each receiving one eighth of them. Each PP module receives data of the same set of events from 5 TSO modules and merges them together. This distributing-merging process partially builds the event as in dedicated data switch equipments. Similar distributing–merging functions are also performed between the PP modules and the Segment Trackers (ST).

The time-stamp ordering FPGA on the TSO module is shown in Figure 6.23. The inputs are 3 channels of serial links at 2.5 Gb/s, which are deserialized into parallel data. The input data are stored in two zero turn-around (ZBT) synchronous RAM devices and waited until data from a beam cross-over (BCO) are believed to have all arrived.

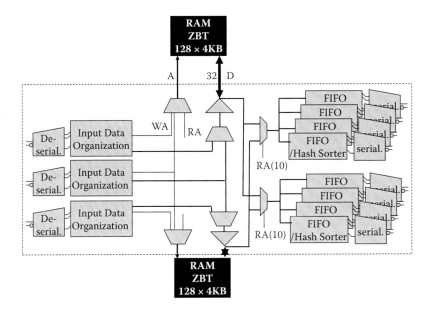

Figure 6.23 Parasitic event building in an FPGA.

Then the data from a BCO are read back into the FPGA and sent into one of the 8 output channels. Data for next BCO are given to next channel, and so on.

The primary function of the FPGA is to perform the given preprocessing function, while it also serves as a data merger and distributor. In other words, it performs a switch fabric function parasitically.

6.10   Digital phase follower

Serial communication is a popular data transmission scheme since the communication channel is simply a twist-pair of a cable. There are serial transceivers available in several FPGA families with an operating bit rate higher than 1 Gb/s. However, the costs of the FPGA with built-in transceivers are normally higher than the comparable devices without transceivers. There are applications of serial communications to be implemented in low-cost FPGA families where there are no dedicated transceivers, but relatively lower bit rates are sufficient. The Digital Phase Follower (DPF) is developed to fulfill these needs.

The transmitter of serial data is relatively simple and is parallel to a serial converter implemented either with a shift register or dual-port memory with single-bit output port. The receiver is more complicated since the cable delay causes the data to arrive at any possible phases. When the temperature of the cable varies, the phase of the serial data may drift away from the original phase. If the clocks of the sender and the receiver are not derived from the same source, a continuous and indefinite phase drift is expected.

The DPF uses multiple samples of the data stream to detect and to keep track of the input data phase. In each bit time, the input data is sampled 4 times at 0, 90, 180, and 270 degrees. The 4 clocks (or 0 and 90 degrees, plus their inverted versions) can be generated with a PLL block now available in most low-cost FPGA families. The block diagram of the DPF is shown in Figure 6.24a.

The multisampling part is the same as in the FPGA TDC discussed before and, in fact, the operation of the DPF is based on the transition time of the input data stream. After multisampling, the sampled pattern is first converted to the 0° clock domain. Then 7 samples, QD to Q3, are sent to the transition detection logic to find the relative phase of the input data as shown in Figure 6.25. Note that the sample pattern jumps up 4 bits every

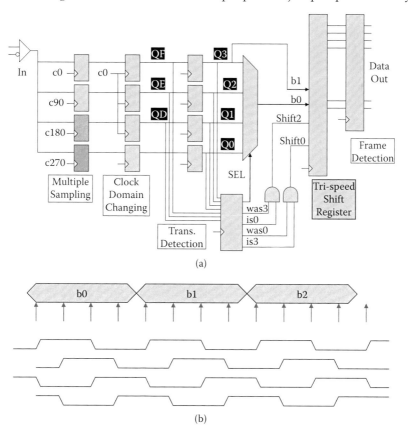

Figure 6.24 The multisampling and digital phase follower: (a) the block diagram, (b) data bit sampling points.

Older Samples

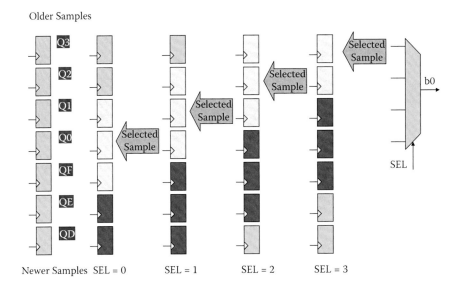

Newer Samples SEL = 0   SEL = 1   SEL = 2   SEL = 3

Figure 6.25 Transition detection in the digital phase follower.

clock cycle; that is, QD jumps to Q1, QE to Q2, and QF to Q3, which is obvious from the pipeline structure shown in Figure 6.24.

The designers may choose to detect both 0-to-1 and 1-to-0 transitions, but detecting only one transition is recommended since the rising and falling time of the input circuit may be different. Once the first transition is seen, its location is registered, and the data sample sufficiently far away from the transition points is selected as an input of the shift register in the later stage. For example, when a 0-to-1 transition is seen between QF and QE, that is, (QF==0)&&(QE==1), the sample Q0 is selected, and so on, as shown in Figure 6.25. When the phase of the input data drifts away from the original point, the transition at a different location is detected. The sample point being selected follows the change of the transition location change accordingly.

As mentioned earlier, the input data phase may drift indefinitely if the clocks of the sender and receiver have very close but slightly different frequencies. Even in the systems with same clock source for the sender and receiver, the phase drift due to cable temperature variation may also be bigger than a bit time.

We can assume that the phase drift rate is not too high so that the position of the current transition is either the same as what was previously detected, or +1 or −1 from the previous position. Under this assumption, there are two possible cases for the bit phase drifting out of one bit time, that is, "was-0-is-3" and "was-3-is-0" as shown in Figure 6.26.

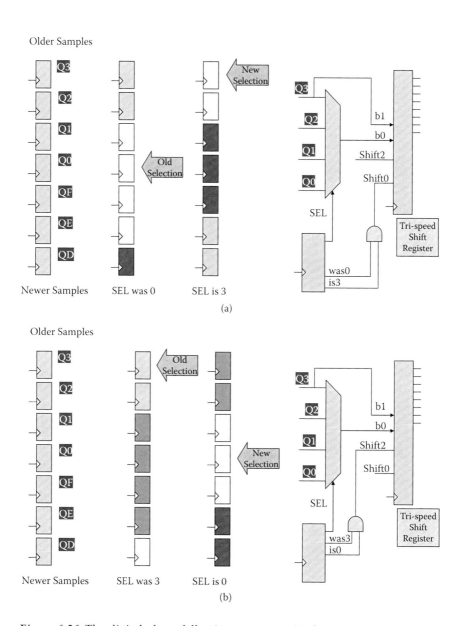

Figure 6.26 The digital phase following processes: (a) the "was-0-is-3" case, (b) the "was-3-is-0" case.

The first case, "was-0-is-3," is finding the transition between Q2 and Q1 requesting that the sample Q3 being selected in current clock cycle while the previously registered selection was Q0. This "was-0-is-3" case indicates that the input data clock is slower than the local clock. The selection point has actually drifted from Q0 to QF. However, to prevent the sampling point from drifting down indefinitely, it is wrapped over to Q3 instead of QF. Since the current sample at Q3 has been shifted into the shift register in the previous clock cycle (which was Q0), the shift register stops shifting for one clock cycle to compensate for the slower input data clock.

The second case, "was-3-is-0," is finding the transition between QF and QE, requesting that the sample Q0 be selected in the current clock cycle, while the previously registered selection was Q3. This "was-3-is-0" case indicates that the input data clock is faster than the local clock. The selection point has actually drifted from Q3 to Q4 (which is not implemented). However, to prevent the sampling point from drifting up indefinitely, it is wrapped over to Q0. In this situation, two sample points, that is, Q3 and Q0, must be pushed into the shift register, causing it to shift by 2 bits in the current clock cycle to compensate for the faster input data.

The shift register normally shifts by 1, and it shifts by 0 or by 2 in the "was-0-is-3" or "was-3-is-0" cases, respectively, which is why it is called the "tri-speed shift register." Typical deserialization circuits, either in the FPGA or a single IC chip, recover the receiving clock using either PLL or clock swapping schemes. The digital phase follower is a pure digital circuit, and the clock recovery is avoided.

The digital phase follower described previously is intended for a relatively high data rate; that is, a bit time is the same as the period of the FPGA internal clock. If the required data rate is lower so that a bit time contains two or more cycles of the FPGA internal clocks (for example, the data rate is 200 M bits/s and the internal clock frequency is 400 MHz), a simpler receiving circuit using even less silicon resources can be designed. In this case, the shift register will not need to shift by 2. The decoder functions in the later stage will be simpler.

6.11   Multichannel deserialization

Typical serial-to-parallel conversion uses a shift register plus a hold register scheme as shown in Figure 6.27a. For data concentration applications where many serial channels are to be converted to parallel words and merged together, a relatively large amount of logic cells may be needed. Use the Fermilab BTeV pixel readout system as an example where each channel outputs a bit stream representing 24-bit words and up to 72 channels are to be merged together.

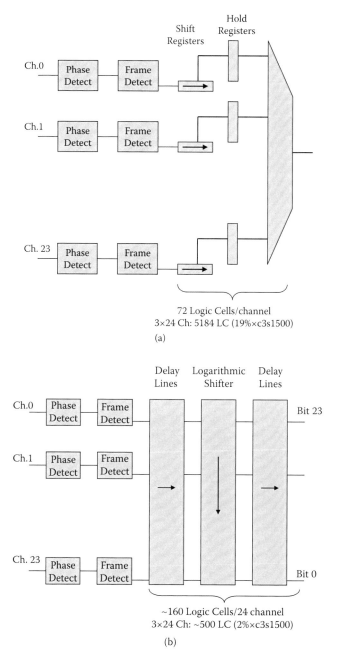

Figure 6.27 The multichannel deserialization schemes: (a) typical scheme, (b) delay-shift-delay scheme.

If the data channels are driven by the same clock source and cable delay variation is expected at less than several bit times, a deserialization scheme, the delay-shift-delay scheme as shown in Figure 6.27b, with a relatively small resource usage can be applied.

The delay-shift-delay scheme takes advantage of SRL16 primitives found in Xilinx FPGA devices. The SRL16 primitive is actually the 16 × 1-bit RAM used for the 4-input lookup table in each logic cell. The RAM can be configured as a 16-step serial-in-serial-out shift register that can be used as a delay line, which would need 16 logic cells if implemented otherwise. See Reference 11 for a detailed description of SRL16.

The delay-shift-delay scheme is developed based on Reference 12 with modifications. All serial inputs are checked for the transition phase using schemes such as the digital phase follower described earlier. The clocks driving all channels are derived from the same source, so the phases of inputs will not drift indefinitely, and the variation of the cable delay is expected to be less than a certain number of bit times. Therefore, the wrap-over and tri-speed shifting processes in DPF will not be necessary with an extended sample pattern. The input frame, that is, the first bit of the first word when a data transfer is started is detected.

The delay-shift-delay scheme is explained in Figure 6.28 and we use converting 8 channels of 8-bit word as the example for clarity. The input stream with the first bit known for each channel is fed into a delay line implemented with SRL16 primitives. The lengths of the delay lines for different channels are adjusted so that channel 0 has shortest length, and channel 7 has the longest with a 1-step difference between adjacent channels. The bit streams of inputs to and outputs from the first delay line stage are as shown in the top row of Figure 6.28.

The bit streams then are sent into a logarithmic shifter stage. The shifter rotates the bit pattern in each clock cycle by 7, 6, 5 … 0 bits, and the

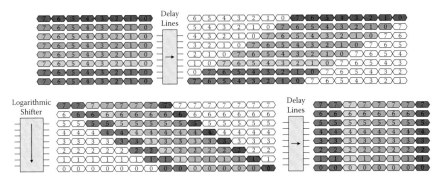

Figure 6.28 (See color insert following page 82.) The delay-shift-delay multi-channel deserialization scheme.

outputs are as shown. After the shifter, all the bits are at the correct output streams but the bits of the same word, for example, data from a channel, are at different clock cycles.

The second delay line stage is used to adjust the output streams. The delay line for bit 7 has the shortest length, while bit 0 has the longest, with a 1-step difference between the two adjacent bit-streams. The final output streams have correctly aligned parallel data words from channel 0 to channel 7, one channel per clock cycle.

References

1. A. Amiri, A. Khouas, and M. Boukadoum, On the Timing Uncertainty in Delay-Line-Based Time Measurement Applications Targeting FPGAs, in *IEEE International Symposium on Circuits and Systems, 2007*, pp. 3772–3775, May 7–10, 27–30, 2007.
2. J. Song, Q. An, and S. Liu, A High-Resolution Time-To-Digital Converter Implemented in Field-Programmable-Gate-Arrays, in *2005 IEEE Transactions on Nuclear Science*, vol. 53, pp. 236–241.
3. M. Lin, G. Tsai, C. Liu, and S. Chu, FPGA-Based High Area Efficient Time-To-Digital IP Design, in *TENCON 2006. 2006 IEEE Region 10 Conference*, pp. 1–4, November 2006.
4. J. Wu, Z. Shi, and I. Y. Wang, Firmware-Only Implementation of Time-To-Digital Converter (TDC) in Field Programmable Gate Array (FPGA), in *2003 IEEE Nuclear Science Symposium Conference Record*, vol. 1, pp. 177–181, October 19–25 2003.
5. S. S. Junnarkar et al., An FPGA-Based, 12-Channel TDC and Digital Signal Processing Module for the RatCAP Scanner, in *2005 IEEE Nuclear Science Symposium Conference Record*, vol. 2, pp. 919–923, October 23–29, 2005.
6. M. D. Fries and J. J. Williams, High-Precision TDC in an FPGA Using a 192 MHz quadrature clock, in *2002 IEEE Nuclear Science Symposium Conference Record*, vol. 1, pp. 580–584, November 10–16, 2002.
7. J. Wu and Z. Shi, The 10-ps Wave Union TDC: Improving FPGA TDC Resolution beyond Its Cell Delay, in *2008 IEEE Nuclear Science Symposium Conference Record*, pp. 3440–3446, October 19–25, 2008.
8. J. Wu, On-Chip processing for the Wave Union TDC Implemented in FPGA, in *2009 IEEE-NPSS Real Time Conference Record*, pp. 279–282, May 10–15, 2009.
9. J. Wu, An FPGA Wave Union TDC for Time-of-Flight Applications, in *2009 IEEE Nuclear Science Symposium Conference Record*, pp. 299–304, October 25–31, 2009.
10. J. Wu et al., Integrated Upstream Parasitic Event Building Architecture for BTeV Level 1 Pixel Trigger System, in *2006 IEEE Transactions on Nuclear Science*, vol. 53, pp. 1039–1044.
11. Xilinx Inc., Using Look-Up Tables as Shift Registers (SRL16) in Spartan-3 Generation FPGAs, 2010, available via: http://www.xilinx.com/.
12. Xilinx Inc., Serial-to-Parallel Converter, 2004, available via: http://www.xilinx.com/.

chapter seven

Examples of an FPGA in advanced trigger systems

7.1   Trigger primitive creation

The flexibilities that field-programmable gate array (FPGA) devices provide make it possible to build trigger systems that produce advanced event identification abilities only available in software triggers or the offline analysis stage in traditional experiments. We use a time-of-flight (TOF) detector shown in Figure 7.1 as an example, assuming the discriminated logic level signals from photo multiplier tubes (PMT) are sent to an FPGA.

At first glance, there is not much that can be done in an FPGA. In the time domain, the signal arrival time depends on the velocity of the particle and the position of the hit on the TOF counter. In space, the hits of particles from the two-body decay are not separated 180° "back-to-back." The separation angle depends on the decay momentum and polar angle of the tracks.

Traditionally, input hit signals are stretched to cover the latest arrivals. Using stretched signals, some approximate coincidences, probably covering a wide range of azimuth angles, can be formed.

However, if the arrival times of the input signals are digitized using time-to-digit converter (TDC) schemes presented in Chapter 6 to a precision of about 0.5 ns, interesting features of the events in several aspects can be extracted.

Assuming the photo multiplier tube (PMT) signal arrival times from both ends of a counter are T_A and T_B, respectively, then the time of charged particle hitting the counter T_H and the hit position Z_H can be approximately found by simply adding and subtracting two arrival times:

$$T_H = (T_A + T_B) / 2 + T_C$$

$$Z_H = C (T_A - T_B) \tag{7.1}$$

The parameter T_C and C are known constants for the given size of the time-of-flight (TOF) counter. The z-measurement can be good to about 10 cm with TDC of 0.5 ns precision or better. With z-positions of charged

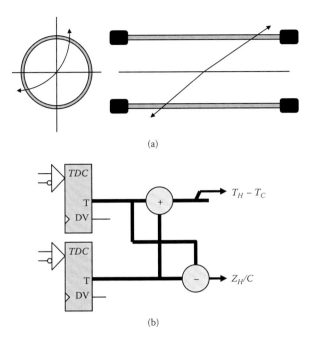

(a)

(b)

Figure 7.1 Time-of-flight (TOF) (a) detector and (b) trigger primitives.

particle hits, the 3-D feature of the event is available allowing the users to identify the type of the event and provide advanced trigger primitives to the global trigger.

In Figure 7.2, the times-of-flight as functions of hit position Z in a TOF detector with radius 0.8 m and half-length 1.2 m for different v/c values are plotted. With time measurements good to 0.5 ns, one should be able to distinguish a slow from a fast particle hit.

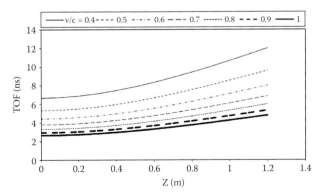

Figure 7.2 The time-of-flight (TOF) as a function of the hit position.

Some backgrounds such as the collision of beam particles with residual gas occur earlier than the fast particles, which can be easily identified.

The *z*-position of the hit itself is very useful, also. It is possible to check the back-to-back condition not only in the *x-y* projection, but also in the *r-z* projection. For example, in a two-body decay event with two opposite-charged particles, once the polar angle of the decay tracks is known, decay momentum can be estimated from the separation of hits in the azimuth dimension. Combining the fast/slow particle identification, the types of the events can be classified fairly well.

The hit positions can also be checked with other detectors in the global trigger stage.

7.2 Unrolling nested-loops, doublet finding

In HEP trigger systems, sometimes data items from two sets are to be paired up. Consider an example shown in Figure 7.3 with two (or more) detector planes detecting tracks coming from a known interaction point chosen as the origin. The magnetic field is in the *y*-direction so that the *x-z* is the bend view and the *y-z* is the nonbend view. To reconstruct tracks, hits on different detector planes generated by the same track are to be paired up.

In the nonbend view, a constraint (i.e., a necessary condition for two hits to belong to a track) exists. Clearly, the *y*-coordinates of two hits from the same track must satisfy the following condition:

$$\frac{y_{1i}}{Z_1} = \frac{y_{3i}}{Z_3}$$

(7.2)

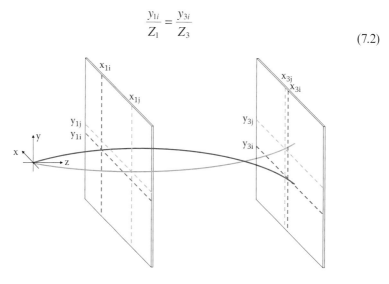

Figure 7.3 (See color insert following page 82.) A hit pairing problem.

The parameters Z_1 and Z_3 are known constants, representing the distances of the two detector planes to the interaction point.

In practice, the constraint is an inequality to allow errors ε from either coordinate measurement or the model itself:

$$y_{1i} + \varepsilon > \frac{Z_1}{Z_3} y_{3i} > y_{1i} - \varepsilon$$

(7.3)

In software, pairing up two data items is implemented with two layers of nested loops, assuming in an event, there are $n1$ and $n3$ hits on the two detector planes:

```
for(i1=0; i1<n1; i1++){
    for(i3=0; i3<n3; i3++){
        CheckCondition(y1[i1], y3[i3]);
    }
}
```

The execution time of the process is $O(n^2)$, where n is the number of hits on the two planes in an event.

In an online trigger system, the process time must cope with the data-fetching time, which is proportional to the number of hits on a detector plane, that is, $O(n)$. It is then necessary to reduce the execution time of the process by "unrolling" a layer of the nested loops.

7.2.1   Functional block arrays

In principle, it is possible to implement multiple copies of a function in the FPGA so that multiple operations of a given function can be performed in each clock cycle. Since multiple copies are created, the functional block array consumes silicon resources rapidly. To reduce silicon usage, the function should be analyzed, and operations that can be done outside the array should be extracted as much as possible.

In the hit matching process, for example, hit coordinates from two different detector planes are to be paired up. It is possible to design a functional block array as shown in Figure 7.4. The hit coordinates from one plane are first loaded into the array and are stored in u_1, u_2, etc. (corresponding to y_{3i}, y_{3j}, etc.). Then, the hit coordinate x (corresponding to y_{1i}, y_{1j}, etc.) from another plane for each hit is fetched in, and all functional blocks check the matching conditions with prestored values u_1, u_2, etc., simultaneously. The process repeats for the next hit coordinate x until all hits are looped through.

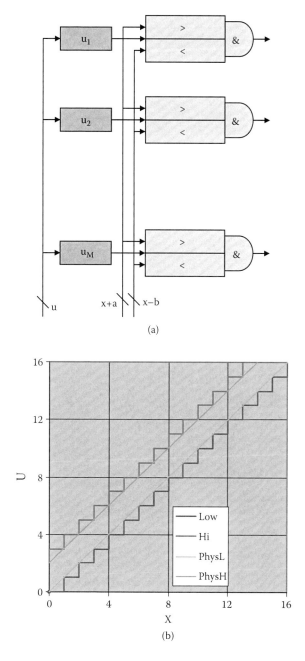

Figure 7.4 Range cutting (a) functional block array; (b) acceptance map.

It is possible to extract the operations common for all blocks, leaving two comparison operations in each functional block. The hit coordinates from one plane are first scaled and stored in u_1, u_2, etc. The coordinates from another plane are then added and subtracted with two constants, a and b, representing errors in the hit matching constraint. For each x, the prestored values u_1, u_2, etc., are compared with the upper and lower boundaries $(x + a)$ and $(x - b)$, respectively. If a prestored value falls inside the boundary, two hits possibly forming a track are paired up together.

If there are m and n hits/event in each detector plane, then the execution time for the whole process takes about $(m + n)$ clock cycles, if $m < M$, where M is the number of blocks implemented in the array. The execution time of the process is then reduced from $O(n^2)$ to $O(n)$ at the cost of multiple copies of the comparator blocks.

The functional block array above is significantly smaller than if the addition and subtraction operations are performed in each array element. The acceptance map has relatively smooth upper and lower boundaries if sufficient bits are implemented in the comparators.

It is possible to further replace magnitude comparators with equality comparators as shown in Figure 7.5. If implemented with FPGA logic elements, the equality comparator is about half the size of the magnitude comparator of the same number of bits. The trade-off is, however, that the boundaries in the acceptance map become a relative rough saw-teeth shape.

A functional block array with M block checks up to M pairs of data simultaneously. It "unrolls" the innermost loop if the number of hits m in the inner most loop is $<M$. If $m > M$, multiple passes of operation will be needed.

7.2.2   Content-addressable memory (CAM)

The content-addressable memory (CAM) can be viewed as a functional block array checking single equality as shown in Figure 7.6. The real CAM uses only two additional transistors per memory cell, so it is very resource-conserving, given the complexity of the function that CAM performs.

In FPGA, however, not so many families have the real CAM resources, and it takes a large amount of silicon resources if "implemented" with logic elements. If a large amount of data items with large word width are to be checked, an external CAM device can be considered as an option.

The CAM checks single equality between the input data and the prestored data. An obvious issue is the missing boundary area in the acceptance map. To cover the missing boundary, one may consider using two CAM devices to check for double equality. If there are several clock cycles available for each write and/or read operation, several slightly varied data can be stored in the CAM, or several comparisons can be performed during the read to cover the missing boundary areas.

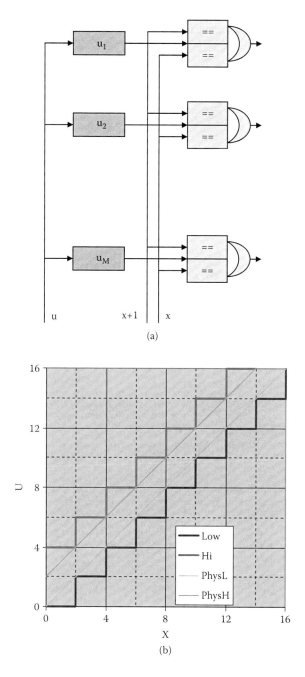

Figure 7.5 Multiequality (a) functional block array; (b) acceptance map.

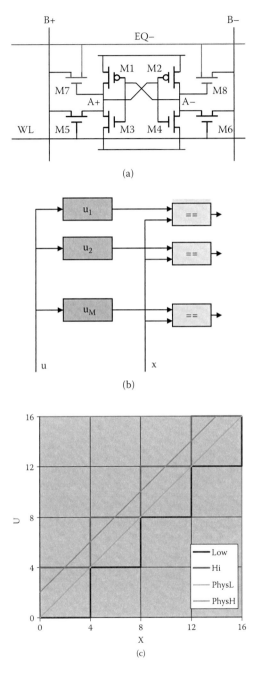

Figure 7.6 The contents-addressable memory: (a) the CAM cell, (b) the single equality functional block array, and (c) its acceptance map.

Typically, the CAM devices are designed to make single matches, such as finding a unique IP address in network equipments. Special care must be taken to handle multiple matching situations, which are very common in HEP applications.

7.2.3 Hash sorter

The hash sorter is a better choice for matching data items in HEP applications in terms of silicon resource usage, multiple data items handling, etc. The hash sorter is implemented using block RAM and logic element resources found in any FPGA. See References 1 and 2 for details.

The hash sorter can be viewed as memories organized into bins that are indexed by a key number K as shown in Figure 7.7. The data items with a particular key number are stored in the particular bin. The data items can be retrieved quickly later.

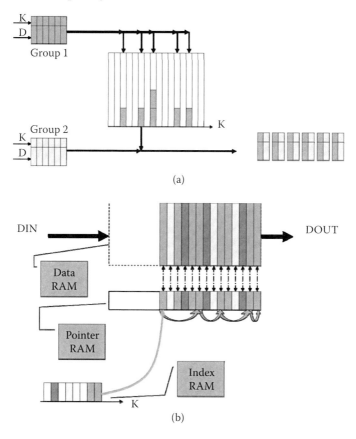

Figure 7.7 The hash sorter: (a) the hash-sorting function, (b) the link list.

Use the track recognition in a nonbend view as an example. The detector plane is divided into, for example, 256 bins. The hit data in one plane are first stored into the hash sorter with the upper 8 bits of the coordinates as the key (bin) number K. Then the hits in another plane are looped over. For each hit in the second plane, the coordinate is scaled as the bin number and is used to address the hash sorter. The addressed bin contains the candidate hit in the first plane that forms a valid track with the hit in the second plane.

In a hash sorter, multiple data items can be stored in a bin. If the memory were equally divided into bins, there would be an upper limit to the number of data items that could be stored in a bin. In a real implementation, the data items in each event are stored in a link list structure, and therefore there is no predefined limit per bin. The only limit is that the total number of data items in all bins must be smaller than the size of the memory.

The hash sorter rearranges the data items based on the key number K. The processing time for n data items is $O(n)$—more precisely, n clock cycles to store them and n clock cycles to pair them with the hits in the second plane, plus a one clock cycle to refresh the hash sorter for the next event. The hash sorting should not be confused with regular sorting, which uses $O(n*\log(n))$ process time with arbitrary precision on the key number K. The precision of the key number K in the hash sorter is typically 8–14 bits, which is limited by how many bins can be implemented in the hash sorter—typically several hundreds to several thousands. The data items stored in the same bin are not ordered, which usually is not a problem in many HEP applications. The simplicity and higher speed outweighs the doesn't-matter disadvantage that the hash sorter does less than regular sorters.

There is no global signal to reset a RAM block, and the applications using the hash sorter usually need to be ready for the next event quickly. To refresh functional blocks using RAM for a new event in a single clock cycle, similar design practices such as implementing histograms with a fast resetting ability, discussed earlier, can be utilized.

7.3 Unrolling nested loops, triplet finding

In HEP experiments, objects with two free parameters are very common. Several examples are shown in Figure 7.8. The track segment in the nonbend view (Figure 7.8a) contains offset and slope as the two free parameters [3]. A circular track segment (Figure 7.8b) normally has three free parameters. But if the collision point (beam axis) is known, two free parameters such as the initial angle and curvature are sufficient to describe a circular track. A hit on a multiwire chamber plane is specified with two coordinates, and a track passing a MICROMEGAS-based time-projection chamber (TPC) pair in the drift direction is described with the x coordinate and track hit time t_0.

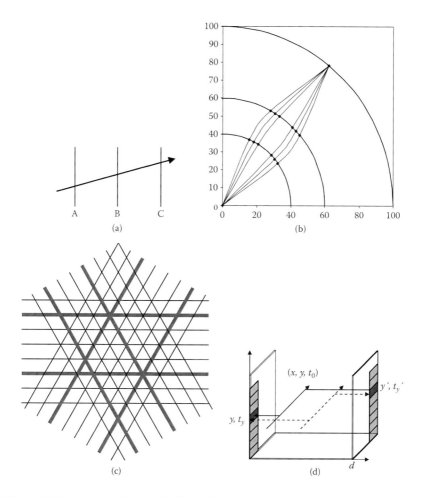

Figure 7.8 Examples of triplet finding: (a) straight tracks in free space (b) circular tracks from the detector center (c) hits on the multiwire chamber (d) tracks passing through a time-projection chamber.

To recognize a possible candidate of the object with two free parameters, at least three measurements must be considered simultaneously. Therefore, objects with two free parameters are also called "triplets." From three independent measurements, the two parameters can be evaluated, whereas an extra constraint is introduced as a necessary condition that the triplet must satisfy. The triplet-finding process uses the constraint to group three data items that form a possible triplet.

In the multiwire chamber example, assume two charged particles hitting different locations that fire two wires in each view. There are 12 intersections of fired wires if only two views are considered. By considering all

three views, the locations of the two hits can be identified as the intersections with fired wires of all three views.

In the MICROMEGAS TPC pair, the coordinate x and hit time t_0 can
be calculated if there is only one pulse within a sufficiently long-time
window in each of the two pads on opposite sides. If multiple pulses are
generated by multiple tracks in a pad pair, without additional measurement, the pulses may be mismatched resulting in ghost hits. A possible
additional measurement can be particle hit times, either derived from an
accelerator RF or measured from other detectors. The additional measurements can also be made with an additional TPC with different drift direc-
tions or different drift velocities.

In software, triplet finding is implemented with three-layer nested
loops, assuming in an event, there are $n1$, $n2$, and $n3$ hits on the three
detector planes:

```
for(i1=0; i1<n1; i1++){
      for(i2=0; i2<n2; i2++){
            for(i3=0; i3<n3; i3++){
                  CheckCondition(y1[i1], y2[i2],
y3[i3]);
                  }
            }
}
```

The execution time of the process is $O(n^3)$, where n is the number of hits on
the three planes in an event.

In an online trigger system, the process time must cope with the data
fetching time, which is proportional to the number of hits on a detector
plane, that is, $O(n)$. It is then necessary to reduce the execution time of the
process by "unrolling" two layers of the nested loops.

Obviously, it is possible to go through a doublet finding stage first and
then find the triplets. However, there are methods to find the triplet directly
without constructing doublets. Several schemes are now discussed.

7.3.1   The Hough transform

Consider the straight-track segment-finding problem in the nonbend
view with the plane A and C being divided into N bins each. A bin can
be a natural detector element, or a group of the elements. There are
a total N^2 to $2N^2$ possible track configurations, depending on how the
plane B is divided. A valid track configuration is called a "road" or a
"pattern."

The track segment is determined by two free parameters. The method
of the Hough transform [4] is to book a multidimensional histogram in the

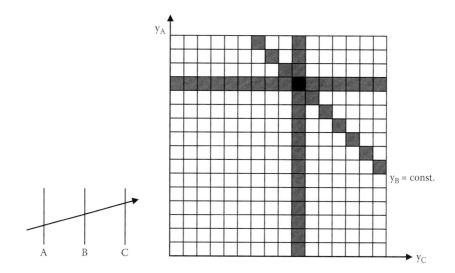

Figure 7.9 Hough transform for triplet finding.

parameter space. (The actual hits are in "image space.") In our example, there are two free parameters, so the histogram is 2-dimensional. A bin in the parameter space represents a set of parameters that correspond to a road. A hit in the image space may be a part of many possible roads that, in general, make a curve in the 2-D histogram. While processing an event, each hit causes a set of cells on the curve to increment by 1. The curves of hits from different detector planes intersect at common cells. After hits from all detector planes are processed, the peaks in the 2-D histogram represent the actual roads of the tracks. The method is illustrated in Figure 7.9.

There are many choices of parameters, and the coordinates of the hits on plane A and C can be chosen as the two parameters for convenience. The 2-D histogram contains N^2 cells; each can be visualized as a counter with count enable input.

A hit in plane A represents a curve in the parameter space, which is a horizontal straight line due to our choice of parameters. Each hit in plane A enables a row of the counters so that they all are incremented by 1. Similarly, a plane C hit enables a column of the counters. A hit in plane B represents a diagonal straight line and enables corresponding counters for incrementing. The intersection of three lines represents the roads of the track. The cell is accumulated to 3 while other cells are not. If there are more detector planes, the hits from different planes represent lines with different angles, and they should all intersect at a common cell if they are created by the same charged particle track. The cell should have a count equal to the number of detector planes. With more than 3 planes, the designer may use more flexible peak-finding algorithms to accommodate detector inefficiency or a boundary effect. For example, with 5 detector planes, a valid peak can be defined as a cell equals 4 or 5, allowing one missing hit.

The obvious drawback is the large $O(N^2)$ silicon usage of the 2-D histogram. Note that each bin or cell in the histogram contains a counter, count enable logic, and peak recognition logic that consume several logic elements in the FPGA.

7.3.2   The tiny triplet finder (TTF)

In many applications, constraint of the triplet is an invariant under shift, that is, the constraint remains valid if all the measurements are added with a constant offset. In the straight-track segment-finding, multiwire chamber and TPC examples, their constraints are linear so they are obviously invariants under shift. The constraint for a circular track is not linear, but it remains valid under rotation, that is, all measured azimuth angles can be added with a common offset. If the application satisfies the shift invariant condition, the tiny triplet finder (TTF) is a suitable scheme for the triplet-finding process. The circular track-segment finding using TTF is shown in Figure 7.10.

The key feature of TTF is its low silicon resource usage: $O(N*\log N)$. The TTF uses logarithmic shifters with silicon resource usage $O(N*\log N)$ to shift hit patterns before feeding a bit-wise coincidence logic so that the coincidence logic is properly reused each clock cycle. The principle of TTF has been discussed in References 5–7.

7.4   Track fitter

In high-energy physics-experiment detectors, track fitting is normally considered a software task in the higher-level trigger stage or analysis stage. Almost all relatively decent fitting algorithms require floating point multiplications and divisions. Although directly porting the fitting algorithm into today's large-size FPGA is not impossible, cost and power consumption quickly become concerns without careful resource usage control. In fact, many silicon area and power consuming operations, such as multiplications and divisions in many algorithms, can be eliminated or replaced by low resource usage operations such as shifts, additions, and subtractions. A process deviating from the mathematically accurate one certainly produces less perfect results. However, significant reduction in FPGA logic elements and power consumption overweighs minor imperfections.

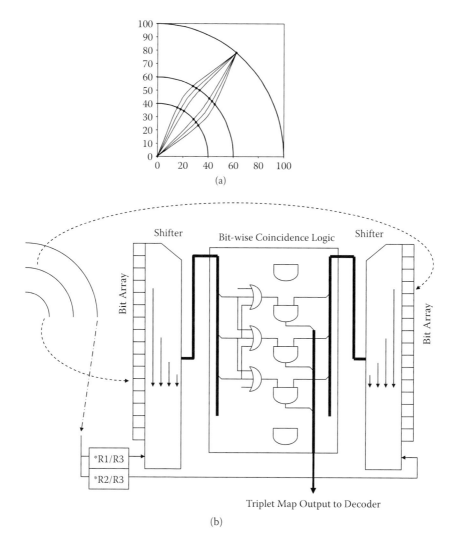

Figure 7.10 The tiny triplet finder (a) curved tracks in a magnetic field (b) block diagram. .

In References 7 and 8, a curved track fitting (as shown in Figure 7.11) algorithm that needs only shifts, additions, and subtractions is analyzed. It is modified from the least-squares fitting method, and its fitting errors are only slightly increased from the mathematically perfect ones.

The five parameters of a track (two offsets, two slopes at the middle of the track, and a curvature in the bend view) are found with the following linear combinations:

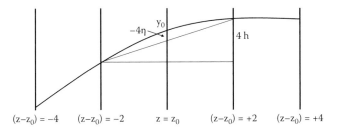

Figure 7.11 The curved track fitting.

$$x_0 = \sum_i a_i x_i / \sum_i a_i \quad l = \sum_i b_i x_i / \sum_i b_i (z_i - z_0)$$

$$y_0 = \sum_i c_i y_i / \sum_i c_i \quad h = \sum_i d_i y_i / \sum_i d_i (z_i - z_0)$$

$$\eta = \sum_i e_i y_i / \sum_i e_i (z_i - z_0)^2$$

$$(7.4)$$

Usually, floating-point multiplication and divisions are needed. However, the coefficients in the linear combinations are modified from the least-squares fitting algorithm. Only "weigh-two" or "two-bit" integers are chosen as the coefficients. An example of choosing e[i], the coefficients for calculating parameter η guided by e_i, the one derived from the least-squares algorithm is shown in Table 7.1.

Since the coefficients are simply two-bit integers, a multiplication of the coefficient is reduced to two shift and accumulation operations. (See Reference 8 for details.)

Since the operations of the algorithm are very simple, it can be implemented into an FPGA with very low silicon resource usage. The track-fitting functional block calculates 5 parameters (two offsets, two slopes at middle of the track, and curvature in the bend view) of a track while the coordinate data of the hits on the detector planes are flowing through. The tracks are allowed to have various numbers of hits so that the advantages of redundant measurements and a long lever arm for tracks with more hits are fully taken. The fitting errors are only increased slightly from the one of the least-squares algorithm (<4%).

It is often the case that in a computation problem, a large percentage of the arithmetic logic operation is spent to make just a few percentages' worth of improvement. With some care, it is possible to eliminate those excessive operations without significantly degrading the final results.

Table 7.1 The Coefficients of the FPGA-Fitting Algorithm

$z-z_0$	Half-Length of the Track													
	16		14		12		10		8		5		4	
	c_i	e[i]	c_i	e[i]	c_i	e[i]	c_i	e[i]	c_i	e[i]	c_i	e[i]	c_i	e[i]
-16	5.3	6												
-14	3.3	2	7.5	8										
-12	1.6	2	4.3	4	11.3	12								
-10	0.1	0	1.6	2	5.6	5	17.9	18						
-8	-1.1	0	-0.7	-2	1.0	1	7.2	7	31.0	31				
-6	-2.0	-3	-2.4	-2	-2.6	-4	-1.2	-1	7.8	8	61.0	56		
-4	-2.6	-3	-3.6	-5	-5.1	-5	-7.2	-8	-8.9	-9	0.0	12	146.3	144
-2	-3.0	-3	-4.4	-4	-6.6	-5	-10.7	-9	-18.8	-20	-36.6	-40	-73.1	-64
0	-3.2	-2	-4.6	-2	-7.2	-8	-11.9	-14	-22.2	-20	-48.8	-56	-146.3	-160
2	-3.0	-3	-4.4	-4	-6.6	-5	-10.7	-9	-18.8	-20	-36.6	-40	-73.1	-64
4	-2.6	-3	-3.6	-5	-5.1	-5	-7.2	-8	-8.9	-9	0.0	12	146.3	144
6	-2.0	-3	-2.4	-2	-2.6	-4	-1.2	-1	7.8	8	61.0	56		
8	-1.1	0	-0.7	-2	1.0	1	7.2	7	31.0	31				
10	0.1	0	1.6	2	5.6	5	17.9	18						
12	1.6	2	4.3	4	11.3	12								
14	3.3	2	7.5	8										
16	5.3	6												
Error	2.91	3.02	3.05	3.15	3.22	3.26	3.41	3.43	3.65	3.65	3.93	3.99	4.28	4.29
Ratio		1.04		1.03		1.01		1.00		1.00		1.02		1.00

References

1. J. Wu, M. Wang, E. Gottschalk, G. Cancelo, and V. Pavlicek, Hash Sorter: Firmware Implementation and an Application for the Fermilab BTeV level 1 Trigger System, Nuclear Science Symposium Conference Record, 2003 IEEE, vol. 2, pp. 1254–1256, October 19–25, 2003.
2. J. Wu et al., A Pattern Recognition Scheme for Large Curvature Circular Tracks and FPGA Implementation Using Hash Sorter, in *Proc. 10th Workshop Electronics for LHC and Future Experiments*, Boston, MA, p. 397, 2004.
3. E. E. Gottschalk, BTeV Detached Vertex Trigger, *Nucl. Instrum. Meth.* A 473, 167, 2001.
4. R. Fruhwirth et al., *Data Analysis Techniques for High-Energy Physics*, 2nd ed., Cambridge, Cambridge University Press, 2000.
5. J. Wu et al., Tiny Triplet Finder (TTF)—A Track Segment Recognition Scheme and its FPGA Implementation Developed in the BTeV Level 1 Trigger System, in *Proc. 10th Workshop Electronics for LHC and Future Experiments*, Boston, MA, p. 68, 2004.
6. J. Wu et al., The application of Tiny Triplet Finder (TTF) in BTeV Pixel Trigger, *IEEE Trans. Nucl. Sci.* vol. 53 no. 3, pp. 671–676, June 2006.
7. J. Wu, M. Wang, E. Gottschalk, and Z. Shi, Curved Track Segment Finding Using Tiny Triplet Finder (TTF), *Nuclear Science Symposium Conference Record*, 2006 IEEE, vol. 2, pp. 1281–1285, October 29–November 1, 2007.
8. J. Wu et al., FPGA Curved Track Fitters and a Multiplierless Fitter Scheme, *IEEE Trans. Nucl. Sci.* vol. 55, no. 3, pp. 1791–1797, June 2008.

chapter eight

Examples of an FPGA computation

8.1   Pedestal and RMS

The pedestal and RMS of an input stream provide commonly used characterizations of the signal sequence. The block diagrams for calculations of the pedestal and RMS are shown in Figure 8.1.

The mean and the standard deviation (squared) of the input stream can be written:

$$\overline{x} = \frac{\displaystyle\sum_{i=0}^{N-1} x_i}{N} \qquad \sigma^2 = \overline{x^2} - \overline{x}^2 = \frac{\displaystyle\sum_{i=0}^{N-1} x_i^2}{N} - \left(\frac{\displaystyle\sum_{i=0}^{N-1} x_i}{N}\right)^2$$

$$(8.1)$$

It is convenient to choose $N = 2^m$, that is, 256, 512, 1024, etc., so that the division can be simply implemented as a bit shift. In fact, the "shift" operation in FPGA usually is just choosing corresponding bits while feeding the data to a later stage, and consumes no real silicon resource.

The circuit of calculating standard deviation uses two accumulators, one to accumulate the raw input data and the other to accumulate the square of each input data. After N data points are accumulated, the bits of sum of the input data are chosen for division to produce the mean. The mean is then squared and subtracted (with appropriate bit alignment) from the accumulator result that represents the sum of squared input data. After the process, the squared standard deviation σ^2 is presented at the output.

To find σ itself, a square root operation can be performed. Perhaps the simplest method for finding the square root is to use a lookup table.

Sometimes the pedestal is relatively large comparing the dynamic range of the noise. For example, in Fermilab BLM system, the pedestal of the ADC input is typically 700 counts while the noise is only 2–3 counts up and down from the pedestal. If the input is squared directly, an input of 10 bits or more would be needed and the result becomes 20 bits or more. The

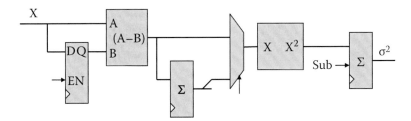

Figure 8.1 Pedestal and RMS calculations.

silicon resource usage of the square operation block with K bits is $O(K^2)$ if the square function is implemented with a multiplier, and is $O(K2^K)$ if implemented using the lookup table. It is known that the standard deviation remains the same if all inputs are subtracted with a common offset. So, in the circuit shown above, the first input data point is chosen to be the offset and all the remaining inputs are subtracted with the offset. The difference stream then contains data with relatively small absolute values, and its dynamic range becomes only 3–4 bits.

Another point that should be mentioned here is that the square operation functional block is reused for both calculating x_i^2 during the accumulation and calculating the square of the mean after the accumulation is done. The square functional block consumes a relatively big silicon resource, so it is a good idea to reuse it as much as possible.

8.2   Center of gravity method of pulse time calculation

In detectors such as calorimeters, particle hits sometimes generate relatively wide pulses, and the ADC is able to sample the pulse to record multiple data points. It is often necessary to determine the arrival time so that the pulse can be tagged with a certain event. The sampling period of the ADC is normally several times bigger than the required precision of the pulse arrival time estimation. For example, the sampling period of a 40 MHz ADC is 25 ns, while the required precision of the arrival time can be 2–3 ns. Time determined by traditional over-threshold discrimination scheme varies, depending on the pulse height and noise. A typical pulse with noise is shown in Figure 8.2.

There are many algorithms for pulse arrival time determination that take advantage of statistical properties of multiple measurements to eliminate the effects of noise and pulse height. The simplest one that can be implemented in the FPGA with low resource usage is the center of gravity scheme as shown in Figure 8.3.

The center of gravity of an input stream x_j is defined as

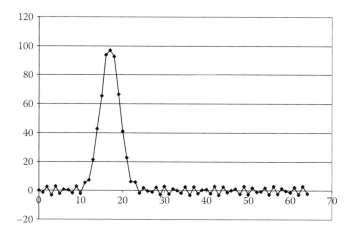

Figure 8.2 A simulated pulse with noise.

$$t = \sum_{j=0}^{N-1} j \cdot x_j \bigg/ \sum_{j=0}^{N-1} x_j$$

(8.2)

It is the ratio of the sum of the input sequence and the sum of the input sequence weighted by the sampling time. Clearly, the pulse height is canceled, and the random noise can be averaged to a relatively lower level in the sums of multiple data points (perhaps a pedestal could be subtracted from the input stream first).

The simple scheme is able to estimate the pulse arrival time accurately to about ¼ to ⅛ of the sampling time interval. The estimated time becomes available shortly after the accumulation of the input sequence covering the pulse is done, so it is good for online coarse event time estimates in a trigger system.

The multiplier in the circuit can be replaced with a logarithmic shifter that uses lower silicon resources. There are usually a few clock cycles in an FPGA for each input data sample since, in general, analog circuits in ADC cannot be driven as fast as digital circuits in an FPGA. For example, for each sample in a 40 MHz ADC, there are 4 clock cycles to process each sample point in an FPGA running at 160 MHz. One may take this advantage to use shift and accumulation to implement the multiplication. For example, an 8-bit multiplication can be implemented with adding or subtracting a number (and its shifted versions) 4 times into an accumulator.

The numerator and denominator for the divider in the last stage are the results of the accumulation of multiple points. So the divider has many clock cycles to do the division since the next numerator and denominator are to be accumulated over a long time. A sequential divider is preferred

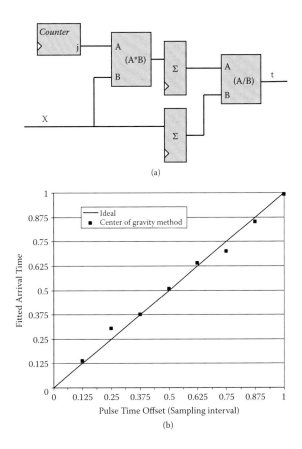

(a)

(b)

Figure 8.3 The center of gravity method for a pulse time estimate: (a) block diagram, (b) fitted result in a simulated example.

due to its compact size to the single-step divider, which uses large logic element resources.

8.3   Lookup table usage

The lookup table is a convenient means of implementing functions. The lookup table is a block memory, usually organized as a ROM. The variable of the function is input as the address, and the prestored content at the address in the ROM is output as the function value.

8.3.1   Resource awareness in lookup table implementation

As pointed out earlier, the silicon resource usage of a lookup table with K bits is $O(K2^K)$, which is a very rapid raise. Based on this fact, the number of bits of the input variable should be limited. As of today, a reasonable width of the block memory address is around 8–12 bits in FPGA. If we trust Moore's law, that everything doubles every 18 months, there are still about 30 years to go before we can easily implement 32-bit lookup tables in the FPGA.

In principle, the functions implemented with lookup tables can be multivariable ones. However, to share 8–12 bits among several variables results in poor precision on each variable. So typically the lookup tables are used to implement single variable functions. In certain cases, for example, to find a ratio of two numbers with 4–5 bits precision, the division can be implemented with a lookup table. Otherwise, to find Y/X, it is a common practice to use a lookup table to find 1/X first, and then use a multiplier to find Y*(1/X). This way, the division with 8–12 bits precision can be performed at a rate of one operation per clock cycle.

8.3.2   An application example

In accelerator physics, simulating the charged particles' movement and positions requires large amounts of computing power. In simulations, determining the magnitude and direction of the particles' movement requires calculating the force using Coulomb's law. The following calculation shows the force on one particle if there are n particles in the system (where r is the position vector and q is the charge):

$$\mathbf{F}_i = \frac{q_i}{4\pi\varepsilon_0} \sum_{j=1}^{n} \frac{q_j \left(\mathbf{r}_i - \mathbf{r}_j\right)}{\left|\mathbf{r}_i - \mathbf{r}_j\right|^3}$$

(8.3)

The same calculation is repeated for all other particles before the positions are updated for each time step. Consequently, the number of total computations is $O(n^2)$, that is, when the number of particle increases by a factor of 10, computation time increases by a factor of 100. A 16-bit demo core for Coulomb forces computing is implemented in an Altera Cyclone II device (EP2C8T144C6). The block diagram of the 16-bit demo core for the space charge simulation is shown in Figure 8.4.

The demo core is designed in pipeline fashion with an internal clock of 200 MHz. The 16-bit coordinates of 256 particles are stored inside the FPGA.

In each clock cycle, a pair of particles marked with "i" and "j" is chosen, and the force between them is calculated through the pipeline. Two counters control the computing sequence by circulating the "i" index one count per clock cycle and then incrementing the "j" index after all i-particles have been traversed through. The coordinates of each pair of the particles are subtracted to find the differences Δx, Δy, and Δz as the

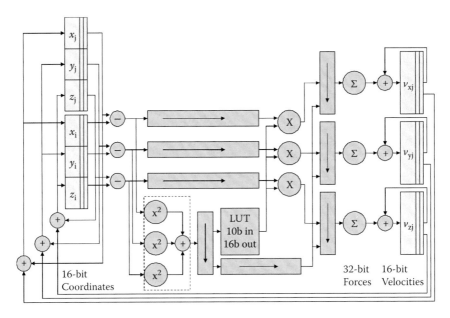

Figure 8.4 Block diagram of the demo core for space charge computing.

components of $(r_j\text{-}r_i)$. The differences are then fed into a sum-of-squares block to calculate the square of the distance between the two particles $|r_j\text{-}r_i|^2$. This value is used to address a lookup table (LUT) of the inverse square-root cubed. The output of the lookup table is then multiplied with the Δx, Δy, and Δz to calculate the component of the force. Note that differences of the coordinates are delayed in a pipeline with the number of stages matching the stages of the sum-of-squares and the LUT. The components of the force on the seed particle marked with "j" are accumulated internally in 32-bit precision.

Because an extensive amount of computing resources and time would be needed to calculate the inverse square-root cubed, the value is calculated by checking a memory lookup table. Instead of using 32-bit r-squared values, we utilize only the nine or ten most significant bits to save memory resources. (If 32 bits were used to address the lookup table, a memory with 4G words would be needed, which would be too large for typical FPGA devices.) This process is performed by removing higher bits of logic 0, and narrowing the value stage-by-stage from 32 bits, to 16 bits, to 12 bits, to 10 bits with a logarithmic shifter. (Sometimes it is called *barrel shifter.* The term *logarithmic shifter* is chosen here to emphasize its implementation scheme and small resource usage.) The 16-bit output value from the table is multiplied by Δx, Δy, and Δz to form force components. Finally, the force components are shifted back to the corrected scale by padding logic 0 bits (or logic 1 in higher bits if the value is negative.). The input for the inverse square-root cubed LUT is shifted downward in increments of 2. If it is shifted downward by $2n$ bits (i.e., divided by $2^{[2n]}$) the inverse square-root cubed LUT output is overscaled by a factor of $2^{(3n)}$. The force components are shifted downward by $3n$ bits to recover back to the corrected scale.

The lookup table is shown in Figure 8.5.

Since the input for the inverse square-root cubed LUT is shifted downward in increments of 2, the address for the LUT is in the range from 256 to 1023 if the raw sum-of-squares is larger than 255. The only possible case for the LUT address to be 0-255 is when the two particles are placed closer than physically possible. In order to account for small r-squared values that would cause the inverse square-root cubed value to approach infinity, output values are limited at the hexadecimal 7FFF. Points with the same x, y, or z position always result with a force of 0 because Δx, Δy, and Δz are multiplied into the table output result.

Since the velocity change in a time step is directly proportional to the acceleration or the force, the accumulated net force components are added together and stored in the 16-bit velocity memories.

The x, y, and z positions are updated after all particles' velocities have been calculated.

The two counters control the entire sequence. One updates each clock cycle while the other updates the particle's new velocity after calculating its net force. After this two-layer nested loop, interactions between all particles are calculated. Then a single loop runs to update position memories, and the iteration for a time step is complete.

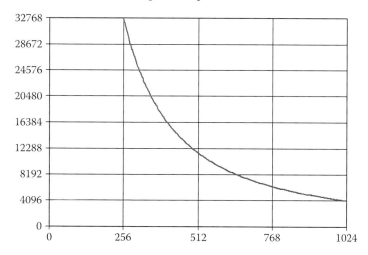

Figure 8.5 The lookup table of the inverse square-root cubed.

In this demo core, the updating of velocity components and position coordinates is performed inside the calculating FPGA as shown above. However, in really large-scale implementations, the accumulated force components are to be read back to the interface FPGA to update the velocity components and coordinates of the seed particle, since this process is only performed once after the interactions between the seed particle and all other particles are calculated.

It should also be pointed out that a quarter of the total memory locations between 0-255 can be used more efficiently. One possibility is to increase input resolution when the input value is 256.0 to 511.5 by adding an extra bit into the LUT address lines.

8.4   The enclosed loop microsequencer (ELMS)

The sequence control of the data processing resources is an important issue in FPGA design.

Sequence control is normally implemented using either finite state machines (FSM) or embedded microprocessor cores. When an input data item is to be fed through a fast and very simple process, typically using a few clock cycles, the FSM is a suitable means of sequence control. FSM also responds to external conditions promptly and accurately. However, the sequence or program in the FSM is not easy to change and debug, especially when irregularities exist in the sequence. Also, the state machines occupy logic elements no matter how rarely they are used. So it is not economical to use the FSM to implement the occasionally used sequences such as initialization, communication channel establishment, etc.

An embedded or external microprocessor is another option of sequence control. Today's mainstream microprocessors are ALU (arithmetic logic unit)-oriented. The ALU, being the centerpiece of the microprocessor, performs not only data processing, but also program control functions. The ALU-oriented architectures have two drawbacks in FPGA computation. (1) When a microprocessor core is embedded in an FPGA, the ALU occupies a large amount of silicon resources. In instances where the application specific data processing is implemented in dedicated logic for the sake of speed, the ALU is barely utilized. (2) The program loops are implemented using conditional branches, which are the primary source of the microcomplexity of pipeline bubble, branch penalty, etc., that needs to be solved with further micro-complexities such as branch prediction. The microprocessor is a better choice only if a data item is to be processed with a very complicated program, typically using thousands of clock cycles.

When a data item is to be processed with a medium length program—for example, using a few hundreds clock cycles—the sequence control needed is not too much more than a program counter (PC)+ROM structure as shown in Figure 8.6a, which is the starting point of the Enclosed

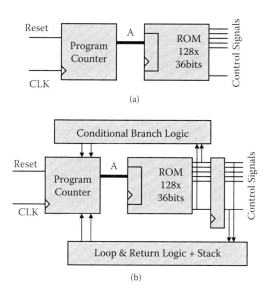

Figure 8.6 The microsequencers; (a) PC+ROM structure, (b) the ELMS.

Loop Micro-Sequencer (ELMS) [1] shown in Figure 8.6b. The primary difference between the ELMS and regular microprocessor is that in the ELMS there are no data processing resources like an ALU. The control signals for external data processing resources turn on and off according to the sequence stored in the ROM as the program counter (PC) increases. Obviously, supporting logic must be added to control the PC. In addition to the conditional branch logic that also exists in microprocessors, loop and return logic with an internal stack are added in the ELMS, so that it supports "FOR" loops with constant iterations at the machine code level and is self-sufficient to run multilayer nested-loop programs.

Most program loops in microprocessors are implemented with conditional branches that are the origin of many microcomplexities such as branch prediction. Intrinsically, loops with predefined iterations need not use conditional branches. The ELMS supports the "FOR" loops with constant iterations at the machine code level, which provides programming convenience and avoids microcomplexities from the beginning. The ELMS is able to run multilayer nested-loop programs without help from external arithmetic/logic resources used for data processing. (In digital signal processor [DSP] devices, a similar feature called "zero overhead looping" [2] is supported in the hardware.) Since the data processing resources are external and purely user defined, the ELMS is not a traditional microprocessor, which is why it is called a "microsequencer." The ELMS is used in the digitizer FPGA for the Fermilab Beam Loss Monitor system with expected performances.

References

1. J. Wu et al., Readout Process and Noise Elimination Firmware for the Fermilab Beam Loss Monitor System, in 2007 *IEEE Real-Time Conference Record,* April 29–May 4, 2007, pp. 1–8.
2. Analog Devices Inc., Application Note, AN-393: Considerations for Selecting a DSP Processor, 2008, available via: http://www.analog.com/.

chapter nine

Radiation issues

The popularity and versatility of FPGAs extends into domains with high radiation fields, that is, military, space, radiobiology, and high energy physics. The dearth and high cost of radiation-hard CPUs makes the FPGA a low-cost and versatile source of almost unlimited compute power in space. In addition, FPGAs offer the ability to distribute compute and DAQ power throughout the instrument, saving mass in terms of cable harnesses and data transmission power. As shown in the following text, this is in spite of very serious radiation damage issues, requiring special hardware and software mitigation. Some of these issues are in common with ASICs and memories, while some are traced directly to the reconfigurability of the FPGAs.

There are special lines of radiation-tolerant FPGAs [1]. Engineering conferences dedicated to radiation effects [2] cover the FPGA-specific issues. In addition, issues of radiation hardening FPGAs are the topic of a "scientific commons," that is, the SEE Consortium [3], which brings together experts from industry, government, and academia to characterize radiation effects and mitigation techniques for reconfigurable FPGAs.

9.1   Radiation effects

Charged ionizing particles and photons cause total ionizing dose (TID) effects, which are cumulative over time. Displacement damage to the crystal lattice by protons and neutrons is more of a problem for sensors, and not for FPGAs. The dominant problems are single-event effects (SEE) due to the deposit of extremely large charges by single heavily ionizing particles such as heavy ions (H.I.).

9.1.1   TID

Modern ICs, including FPGAs, are fabricated in deep-submicron CMOS which is "accidentally" radiation-hard because of the thinness of the gate oxide. The TID limit for radiation-hard FPGAs is about 300 kRad (3 kGy).

9.1.2   SEE effects

Heavy ions can deposit linear energies transfer (LET) energies to the circuit nodes, which generate charges 1000× larger than those from minimum ionizing particles (MIPs), for example, fast protons. These charges can lead to potentially destructive effect like Latch-up (SEL), in which a parasitic transistor in the device gets turned on, leading to a large current surge and catastrophic failure unless the currents are limited by resistors.

A nondestructive single-event effect is the single-event upset (SEU), at which the bit in a circuit node (memory, register) is flipped. The impact of an SEU on the operation of the FPGA varies depending on where the SEU occurs. If it happens in memory, it might simply result in the corruption of stored data, with no consequence on following events. On the other hand, if it occurs within one of the control logic elements, the circuit might get corrupted leading to a Single Event Functional Interrupt (SEFI). The consequence of a SEFI depends on the function of that node in the circuit, but usually the power has to be cycled and/or the configuration has to be reloaded.

Dynamic testing using clocked operation probes for additional upset sensitivities not observable in unswitched devices are mainly single-event transients (SETs).

9.2   FPGA applications with radiation issues

9.2.1   Accelerator-based science

The fluence of high-energetic protons, neutrons, electrons, and photons can exceed the one in space by many orders of magnitude, but radiation shielding is possible outside the detectors, and much of the control and processing electronics using FPGA and CPUs are in shielded areas. Inside detectors with fluences in excess of 10^{15} hadrons/cm^2, ASICs fabricated with radiation-hard technologies (deep-submicron CMOS, special bipolar BJT) are used.

9.2.2   Space

Space radiation affecting electronics are galactic cosmic rays (GCR) consisting of high energetic protons and heavy ions, trapped protons and electrons/positrons in the radiation belts, and solar particles. A requirements document like the one for the Fermi (formerly GLAST) mission [4] in low-earth orbit (LEO) defines the radiation environment of the space flight.

Shielding is difficult since the required mass is an extremely limited commodity on spacecraft. Total shielding is possible only for electrons, and for protons and heavy ions only for the lowest energies with a few hundred mil of Al. In the radiation belts, that is, in the South Atlantic Anomaly (SAA) in low-earth orbit (LEO), trapped protons with typical energies of a few 100' MeV can generate secondary H.I.s inside the

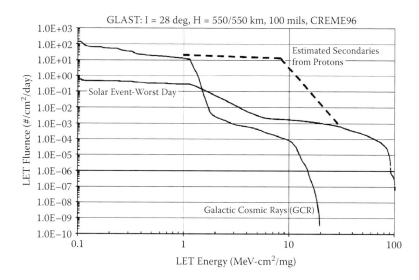

Figure 9.1 Linear energy transfer (LET) spectrum for the GLAST (now Fermi) Mission in LEO. The integrated fluence is calculated for one day. The spectrum for GCR and solar events are from Reference 4 (Science Instrument-Space Craft IRD, Version 0.3 August 3, 1999, http://fermi.gsfc.nasa.gov/science/resources/ao/SI-SC_IRDv.3.pdf), and the estimated secondaries from protons are derived from the proton spectrum in Reference 4 as described in the text.

electronic devices. This "high LET" spectrum of protons has been measured, and in Reference 5, it is shown that this secondary LET spectrum corresponds to about a 10^{-6} fraction of the proton fluence. Figure 9.1 shows for the Fermi mission (formerly GLAST) in LEO [4] the daily integrated LET spectrum for GCR, solar events and an estimate of the secondary effects of protons based on Reference 5. At moderate LET, the proton-induced LET flux can exceed by several orders of magnitude others caused by primary heavy ions. Fermi passes through the SAA approximately every 20 h for about 20 min, so software mitigation is possible after each SAA passage without too much interference with data taking.

In LEO, a typical yearly total ionizing dose is less than a few kRad (few 10's Gy), mainly caused by trapped protons [4].

9.3 SEE rates

The probability of SEE effects is given by the cross section σ_{SEE}, which depends critically on the LET of the particle, which can be either expressed per device (e.g., for SEFI) or per bit (e.g., for SEU in the configuration register or for an memory SEU). The cross-section as a function of LET $\sigma_{SEE}(LET)$ shows typically a LET threshold L_{th}, at which the SEE turns on with a characteristic width W, and a saturation cross-section for the large LET σ_{sat} and is usually fitted with a Weibull function, where S is the power parameter:

$$\sigma_{SEE}(LET) = \sigma_{sat} \bullet \left[1 - \exp(-\{\frac{LET - Lth}{W}\}^S) \right]$$

(9.1)

The rate of SEE R_{SEE} is given by the convolution of the cross-section with the LET spectrum $d\varphi/dLET$ of the particle flux φ (Figure 9.1), multiplied by the number of devices or bits #N:

$$R_{SEE} = \# N \bullet \int_{LET} \sigma_{SEE} \bullet \frac{d\phi}{dLET} dLET$$

(9.2)

We can estimate the SEE error rates for the radiation hardened XILINX Virtex-4 FPGA, which is static random access memory (SRAM)-based. They include configurable logic blocks (CLBs) and block memory (BRAM) modules. The SEU bit error cross sections are 10^{-9}-10^{-8} cm^2 per bit below LET = 10 MeV/(mg/cm^2), depending on the block type [6]. With block size of 15 million bits in the CFG and 5 million bits in the BRAM, the resulting bit error rates are of the order 10–20 per day per device for LEO. Software mitigation needs to be scheduled so that it can keep up with the rate of SEE, and a rule of thumb is that the rate of "scrubbing" should be 10 times faster than the SEU error rate.

The different SEFI types have a device saturation cross-section of σ_{sat} = 10^{-6} cm^2 [6], resulting in an SEFI rate of less than 10^{-4} per day per device for LEO.

9.4   Special advantages and vulnerability of FPGAs in space

FPGAs are attractive since they offer relatively cheap reliable computing power in space, which can be reconfigured during flight. In contrast to ASICs, where the configuration is established by metal traces, the design of the FPGA is established through software residing in the configuration blocks, which can be upset by radiation. In addition, FPGAs are not radiation hard, but are fabricated in radiation-tolerant technologies. The small market for radiation-hard FPGAs limits the technologies used in the fabrication. In ASICs, a SEL can be prevented by using more expensive IC technologies like silicon-on-insulator (SOI), but with the EPI process used in radiation-hardened FPGAs, the SEL immunity is by now very good. ASICs and FPGAs differ in the LET dependence of SEU and SEFI errors. While for ASICs, LET thresholds $L_{th} > 10$ MeV/(mg/cm^2) can be attained that moves the SEE cross-sections out of the high-fluence LET region (cf., Figure 9.1), typical thresholds on FPGAs are below $L_{th} = 1$ MeV/(mg/cm²), where the fluence is relatively high.

9.5   Mitigation of SEU

Since SEUs cannot be eliminated, they have to be mitigated. This is done through redundancy and error detection and correction. These procedures are an integral part of the FPGA hardware and software.

9.5.1   Triple modular redundant (TMR)

Critical blocks in the device configuration or the user's logic are designed in triplicate and their content constantly compared ("polled"). The odd content out is corrected with the content of the two that agree. In traditional FPGAs TMR is implemented using software on a large portion of the device's programmable logic. This process of majority voting, or redundancy, means that two-thirds of the resources, or available logic, is consumed for redundancy and is not available for the user's design.

The Actel RTSX-SU FPGAs use a hardware TMR process with three radiation-hardened flip-flop cells instead of one; so the polling is done on the cell level. In order to protect the configuration, it is stored in "antifuses," and the cells turn out to be especially SEU resistant with LET thresholds of 40 MeV/(mg/cm²) [7].

9.5.2   Scrubbing

During "scrubbing," portions of the configuration memory are overwritten without disrupting operations. The system stays fully operational. Some portions of the configuration memory and interface controls are not able to be scrubbed and therefore still encounter SEU. If a SEFI occurs, the system has to be reconfigured.

9.5.3   Software mitigation: EDAC

For SEU errors, error detection and correction (EDAC) might be possible. Sometimes this is applied to an entire block like RAM, to increase the efficiency. Especially difficult to find and correct are, however, SEU of two bits occurring within one byte of stored data, either from the correlated SEU of adjacent cells from one particle with a very large LET, or from two independent SEU errors within the time window between two scrubs.

Bit errors need to be detected and if possible corrected. In many applications, a parity bit is used to detect bit errors, which has the limitation that one can find out that an odd number of errors has occurred, but not which ones, and so they can't be corrected. More advanced EDAC methods employ a Hamming code, which introduces a system of several parity bits, which allows the determination of the corrupted bits and their correction. For example:

"SECDED" = Single Error Correction, Double Error Detection uses 7 parity bits for a 32 bit word,

"DECTED" = Double Error Correction, Triple Error Detection uses 15 parity bits for a 64 bit word.

9.5.4   Partial reconfiguration

Some FPGA devices allow partial reconfiguration—rewriting of a subset of configuration frames, even during operation—in order to change design behavior without fully reconfiguring a device, or to correct memory upsets in high-radiation environments.

References

1. XILINX http://www.xilinx.com/esp/aerospace.htm, 2009.
2. Radiation and its Effect on Components and Systems (RADECS 2010, Sept. 20-14, 2010, Langenfeld, Austria); IEEE Nuclear and Space Radiation Effects Conference (NSREC 2010, July 19–23, Denver, CO); Military and Aerospace Programmable Logic Devices Conference (ReSpace/MAPLD 2010, Nov. 1-4, 2010, Albuquerque, NM).
3. SEE Consortium, http://www.xilinx.com/esp/aero_def/see.htm.
4. GLAST Science Instrument-Space Craft IRD, Version 0.3 August 3, 1999, http://fermi.gsfc.nasa.gov/science/resources/ao/SI-SC_IRDv.3.pdf.
5. D. M. Hiemstra, and E. W. Blackmore, LET Spectra of Proton Energy Levels from 50 to 500 MeV and Their Effectiveness For Single Event Effects Characterization of Microelectronics, *IEEE Trans. Nucl. Sci.*, vol. 50, December 2006, 2245.
6. G. Allen, G. Swift, and C. Carmichael, VIRTEX-4QV Static SEU Characterization Summary, JPL Publication 08-16 4/08 http://parts.jpl.nasa.gov/docs/NEPP07/NEPP07FPGAv4Static.pdf.
7. Actel Corporation RTSX-SU Application Note Rev. 6, 2010 http://www.actel.com/documents/RTSXSU_DS.pdf.

chapter ten

Time-over-threshold

The embedded particle-tracking silicon microscope (EPTSM)

A good example of a mixed analog–digital FPGA application is the embedded particle tracking silicon microscope, developed originally for investigations in radiation biology [1], but then used extensively for the characterization of radiation effects in silicon strip sensors [2]. It makes use of many FPGA resources, such as the digital clock management (DCM) block, random access memory (RAM), first-in-first-out (FIFO) block, and so on. It was a fertile ground for several student theses at the University of California–Santa Cruz [3–5].

10.1   EPTSM system

The analog parameter is the charge collected on the strips due to the passage of an ionizing particle, and the digital parameters are the addresses of the hit strips, the time of arrival of the hits, and their timing relative to an external trigger signal from a scintillator. Figure 10.1 shows the setup: the beta particles in the beam are counted in the scintillator and are intercepted by the silicon strip detector (SSD). The SSD is read out by a particle microscope front-end ASIC (PMFE), and a commercial XILINX ML405 Embedded FPGA test board [6], read out via the Ethernet into the host computer.

The mix of analog and digital signals is shown in Figure 10.2. The front-end PMFE ASIC integrates the charge from the silicon sensor and compares the output to a threshold voltage. The comparator output is sampled once per read strobe and serialized into the data stream. The number of read strobes for which a comparator output is high is the binary time-over-threshold (TOT). The coexistence of low-noise analog and digital signals (data and clock) on the detector board is made possible by sending the digital data via the low-voltage differential signal (LVDS) lines to the FPGA board. It works because LVDS sends differential current signals as logic ones or zeroes. This causes a net zero current into the ground, which all but eliminates "ground bounce." Because of the limitations on chip area, power, and the number of LVDS drivers and

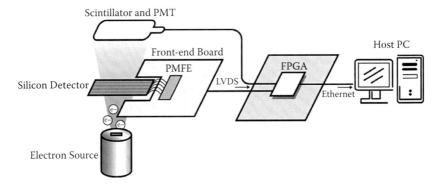

Figure 10.1 Setup of the embedded particle tracking silicon microscope (EPTSM), making use of the XILINX ML405 FPGA board.

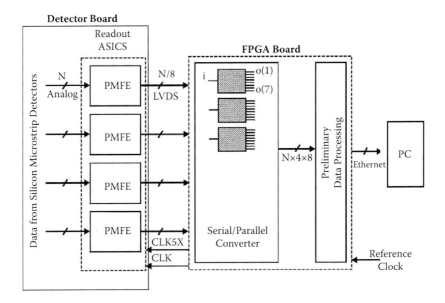

Figure 10.2 Schematic diagram of the PTSM electronics readout. The silicon strip detector (SSD) and the readout ASIC (PMFE) reside on the detector board, connected to the FPGA on the ML405 board via LVDS signals for low-noise operation.

receivers on the FPGA, the data is serialized in the PMFE 8 channels at the time over 8 signal pairs, and deserialized in the FPGA. Note that the data connection from the FPGA board to the host computer consists of a single Ethernet cable.

10.2 Time-over-threshold (TOT): analog ASIC PMFE

The 64-channel analog ASIC PMFE [1] amplifies and shapes to 100 ns the current from the silicon sensor and converts it a voltage. It has a comparator with a variable threshold voltage supplied by the FPGA, which discriminates against noise and outputs a logic signal whose length is correlated with the input charge. The charge Qin collected from silicon detectors scales with detector thickness, and is distributed in the form of a Landau curve, with most events distributed around the most probable value (MPV) from about ½ MPV to several MVP. The typical 300 µm thick detector has MVP = 4 fC, and the threshold is set at about 1 fC, that is, ¼ MPV, to have 100% efficiency even in the case when the charge is shared between adjacent channels.

Figure 10.3 shows a SPICE simulation of the voltage of different input charges Qin at the comparator: the signals are converted into logic signals whose width is the time the voltage levels stay above the indicated threshold. For example, a pulse with charge of one fC barely clears the threshold, resulting in a very small time-over-threshold (TOT), while the pulse of 16 fC gives a TOT of about 3 µs. For signals above 20 fC, the amplifier pulse height saturates, but for higher signals the TOT still grows. Only for signals above 300 fC does the TOT saturate.

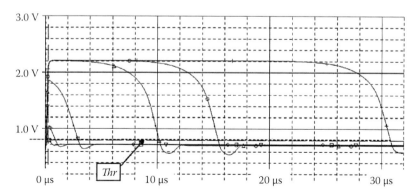

Figure 10.3 Simulated pulse shapes as a function of time at the PMFE comparator for input charges Qin of 1, 4, 16, 64, 100, and 300 fC, respectively. As the pulse height saturates, the pulse length still increases with an increased input charge.

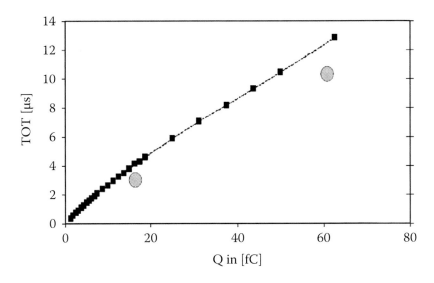

Figure 10.4 Measured time-over-threshold (TOT) in µs versus input charge Qin in fC. The two round symbols indicate the values from the initial SPICE simulation shown in Figure 10.3.

The measured TOT as a function of input charge is shown in Figure 10.4, together with some of the initial simulations extracted from Figure 10.3. The agreement is quite good. As one could guess from the pulse shapes in Figure 10.3, the relationship TOT versus Qin is nonlinear for smaller signals when the pulse height still grows, but becomes linear after the pulse height saturates.

The error in the charge determination σ(Qin) can be determined from the fit of TOT versus input charge Qin (Figure 10.4) by taking the derivative d(TOT)/dQin:

$$\sigma(Qin) = \sigma(TOT) / \frac{d(TOT)}{dQin} \tag{10.1}$$

with the error in the TOT measurement $\sigma(TOT)$ given simply by the reciprocal of the read strobe frequency, multiplied by $\sqrt{2}$ to account for the start and the stop, and divided by $\sqrt{12}$ to account for the equivalent Gaussian RMS:

$$\sigma(TOT) = \frac{1}{10 MHz} \frac{\sqrt{2}}{\sqrt{12}} = 42ns. \tag{10.2}$$

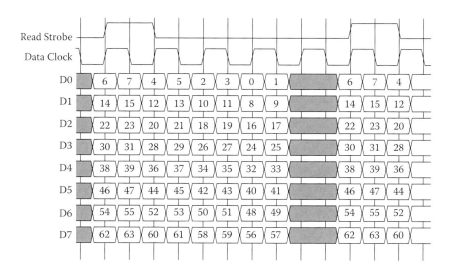

Figure 10.5 Main clock signals and serial data format of a 64-channel PMFE ASIC. Most TOT will span several read strobe cycles ("frames").

For the TOT curve in Figure 10.4, the charge resolution varies between $\sigma(Qin) = 0.14 - 0.22$ fC, a sufficient precision for the approximate width of the Landau curve of $\Delta Qin \approx 4$fC.

10.3   Parallel-to-serial conversion

The comparator outputs are serialized eight channels at a time by clocking them with a double data rate (DDR) into LVDS lines, indicated as D0 ,..., D7 in Figure 10.5. For this, one uses the DCM of the FPGA to send two clocks via LVDS lines to the PMFE, the data clock at 50 MHz, and the read strobe at 10 MHz, which starts a new frame. A TOT in a particular channel is transmitted as valid bits over several frames. Using a DDR and a clock ratio 5 between data clock and read strobe offers 10 potential data bits, of which only the first 8 are filled for the 8 neighboring channels being transmitted (the other resets the sampling latches on the PMFE). The FPGA has 8 pairs of LVDS receivers to clock in the data and deserializes them into 64 parallel channels.

10.4   FPGA function

The FPGA determines the duration of the channel signals (TOT) and their time relationship to other signals. Since the time difference relative to the trigger from the scintillator classifies a channel as noise or valid data, the scintillator signal is piped directly into channel 0 of the FPGA as indicated in Figure 10.1. This allows the trigger decision to be made within the FPGA. The EPTSM is designed to be triggered by any channels, that is, by either the scintillator or by any of the silicon channels, that is, the system can be operated in "self-triggered" mode without a trigger from the scintillator. Since the EPTSM has to be able to simultaneously read out all 64 channels and the scintillator on every clock cycle, it requires a large enough buffer to hold many events on the FPGA before being able to pipe the information along to the controlling computer.

The following functions are performed inside the FPGA:

- Control and calibration
  - Digital clock management (DCM), including shifting the phase of the read strobe to correct for system delays, including cable length
  - Threshold DAC
  - Runtime control
  - Calibration pulse DAQ
- Data handling
  - Serial-to-parallel conversion of data stream
  - Zero suppression
  - Time stamping the start time (+ setting an "up" bit) and stop time ("down" bit)
  - TOT calculation
  - Trigger decision (time coincidence within 3 clock cycles)
  - Formatting: channel # + up/down bit + time stamp
  - Packaging into RAM
- Transmission to on-board CPU

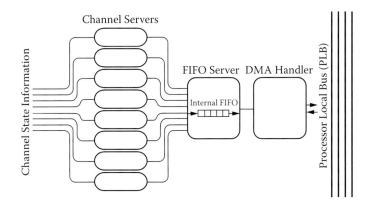

Figure 10.6 FPGA architecture of the EPTSM.

In order to keep up with the data stream, four channels at a time are piped into a "channel server," which buffers and then sends the time stamped data into a FIFO for transmission through DMA via the Processor Local Bus (PLB) to the local CPU (Figure 10.6). The communication between an on-board CPU and its RAM and the host computer proceeds through the Ethernet.

References

1. H. F.-W. Sadrozinski, V. Bashkirov, M. Bruzzi, M. Ebrahimi, J. Feldt, J. Heimann, B. Keeney, F. Martinez-McKinney, D. Menichelli, G. Nelson, G. Nesom, R. W. M. Schulte, A. Seiden, E. Spencer, J. Wray, and L. Zhang, The Particle Tracking Silicon Microscope PTSM. *IEEE Trans. Nucl. Sci.*, vol. 51, pp. 2032–3036, 2004.
2. M. K. Petterson, R. F. Hurley, K. Arya, C. Betancourt, M. Bruzzi, B. Colby, M. Gerling, C. Meyer, J. Pixley, T. Rice, H. F.-W. Sadrozinski, M. Scaringella, J. Bernardini, L. Borrello, F. Fiori, A. Messineo, Determination of the Charge Collection Efficiency in Neutron Irradiated Silicon Detectors, *IEEE Trans. Nucl. Sci.*, vol. 56, pp. 3828–3833, 2009.
3. B. Keeney, The Design, Implementation, and Characterization of a Prototype Read-out System for the Particle Tracking Silicon Microscope, Physics Masters Thesis, University of California–Santa Cruz, 2004.
4. K. Arya, Embedded Particle Tracking Silicon Microscope: An Independent Data Acquisition System for Silicon Detector Characterization, Computer Engineering Bachelors Thesis, University of California–Santa Cruz, 2007.
5. B. Colby, Characterization of Irradiated Silicon Sensors with Time-Over-Threshold, Count Rate and Cluster Size, Physics Senior Thesis, University of California–Santa Cruz, 2008.
6. http://www.xilinx.com/products/boards_kits/virtex4.htm, 2010.

Appendix: Acronyms

2-D: Two-dimensional
3-D: Three-dimensional
ADC: Analog-to-digital converter
ALU: Arithmetic logic unit
ASCII: American Standard Code for Information Interchange
ASIC: Application-specific integrated circuits
BCO: Beam cross-over
β: v/c
BLM: Beam loss monitor
BRAM: Block RAM
BTeV: Name of Fermilab experiment
c: Speed of light
C: (Computer language)
C5: Clock-command combined carrier coding
CAD: Computer-aided design
CAM: Content-addressable memory
CC: Clock and command
CEA: Counter enable
CFG: Configuration block
CIC: Cascaded integrator-comb
CLB: Configurable logic block
CLK: Clock
CM: Centenary mark
CMOS: Complementary metal–oxide semiconductor
CNTEN: Count enable signal
CO: Carry output
CPU: Central processing unit
CRC: Cyclic redundancy check
DAC: Digit-to-analog conversion
DAQ: Data acquisition system
DCM: Digital clock management

DDR: Double data rate
DMA: Direct memory access
DNL: Differential nonlinearity
DPF: Digital phase follower
DQ: Data
DSP: Digital signal processor
DV: Data valid
EDAC: Error detection and correction
ELMS: Enclosed loop microsequencer
EPI: Epitaxial
ε: Error
EPTSM: Embedded particle-tracking silicon microscope
ESB: Embedded system block
f: Frequency
femto: 10^{-15}
fC: Femto Coulomb
FF: Flip-flop
FFT: Fast Fourier transform
FIFO: First-in-first-out
φ: Flux of particles
FPGA: Field-programmable gate array
FNAL: Fermilab, Fermi National Accelerator Laboratory
FSM: Finite state machine
g: Gram
GIGA: 10^9
Gb: Gigabit
GCR: Galactic cosmic rays
GLAST: Gamma-ray large area space telescope (now Fermi Mission)
GHz: Gigahertz
Gy: Gray = 100 rad
HEP: High-energy physics
H.I.: Heavy ion
Hz: Hertz (frequency unit)
IC: Integrated circuit
ID: Identification number
I/O: Input/output
IP: Intellectual property
K: Key (bin) number
k: Kilo (10^3)
L: Length of the pipeline
LAB: Logic array block
LCFF: Logic cell flip-flop
LE: Logic element
LED: Light emitting diode

LEO: Low-earth orbit
LET: Linear energy transfer
LSB: Least significant bit
L_{th}: Threshold LET
LUT: Lookup table
LVDS: Low-voltage differential signal
Mb: Megabit
MEGA: 10^6
mg: Milligram
MHz: Megahertz
micro: 10^{-6}
µs: Microsecond
Micromega: Gaseous particle detector
milli: 10^{-3}
ms: Millisecond
MIP: Minimum ionizing particle
mod: Modulo
MPV: Most probable value
MUX: Multiplexer
N: number of samples
n: Running index
nano: 10^{-9}
ns: Nanosecond
O(...): "to the order of …"
PC: Personal computer, also Program counter
pico: 10^{-12}
ps: Picosecond
PLB : Processor local bus
PLL: Phase-locked loop
PMT: Photo multiplier tube
PMFE: Particle microscope front-end ASIC
PP: Pixel preprocessor
PT: Pulse time
PTSM: Particle tracking silicon microscope
PU: Processing unit
Qin: Input charge
r: Radial coordinate
RA: Read address
Rad: Unit of TID
RAM: Random-access memory
RAW: Read-after-write
RC: Run counter
RE: Read-enabled
RMS: Root mean square

ROM: Read-only memory
R$_{SEE}$: Rate of SEE events
σ: Sigma (RMS error), also cross-section
SAA: South Atlantic anomaly
SCLR: Signal clear
SEE: Single-event effect
SEFI: Single-event functional interrupt
SEL: Single-event latch-up
SET: Single-event transient
SEU: Single-event upset
SOI: Silicon-on-insulator
SRAM: Static RAM
SSD: Silicon strip detector
SSET: Synchronized preset input
ST: Start signal
T: Time
T1: Pulse high time
T2: Pulse low time
TDC: Time-to-digital converter
TID: Total ionizing dose
TMR: Triple modular redundant
TN: Tigger number
TOF: Time-of-flight
TOT: Time-over-threshold
TPC: Time projection chamber
TS: Time stamp
TSO: Time stamp ordering
TTF: Tiny triplet finder
TTL: Transistor-transistor logic
UI: Unit intervals
v: Velocity
VHDL: VHSIC hardware description language
VHSIC: Very-high-speed integrated circuit
W: Width of Weibull curve
WA: Write address
WE: Write enable
x, y: Transverse coordinate
Y2K: Year 2000
z: Longitudinal coordinate
ZBT: Zero turn-around
>>N: Truncated at the Nth bit.

INDEX

A

Accumulator, 5, 6, 13, 24, 47-50, 75, 76, 115
Actel, 129
ADC, 43–47, 50, 53–56, 70–74, 77, 82, 115–117
   see Dithering
   noise, 56
   sampling rate, 43, 44, 54, 116, 117
Adder, 4, 13, 14, 19
Address
   read, 27, 32, 36, 79, 81
   write, 27, 31, 32, 40, 79
Aliasing, 43, 44, 57
   anti-aliasing, 43, 54
Altera, 2, 3, 5, 8, 9, 18, 56, 62, 119
AND, 5, 21, 29, 51, 83
Arithmetic logic unit, 11, 122
ASIC, 1–4, 53, 62, 64, 68, 69, 79, 125, 126, 128,
   131–133
   see Chip

B

Bandwidth, 43, 54, 56, 77
Bit, 4, 5, 7, 12–14, 17, 19, 21- 24, 26, 27, 29, 31,
   36, 38–40, 44, 46–48, 50, 55, 56,
   59–61, 64, 66, 70, 73, 74, 76–78,
   80–83, 85, 88–90, 92–95, 102, 106,
   110–112, 115–123, 126–130, 135,
   136
Bin
   center, 66
   width, 60, 62, 63, 65, 66, 68, 69
Booking
   histogram, 33–37
   session, 35, 36
Buffer
   Inverting, 63, 69
   non-inverting, 63
   tri-state, 14, 15
Burst, 69–70

Bus, 13, 31, 38, 39, 136, 137
   tri-state, 14, 15

C

Cable, 8, 60, 70, 82, 86, 88, 90, 94, 125, 133,
   136
Calibration, 31, 62–67, 136
Carry, 4
   chain, 5, 6, 17, 19, 62–64, 67-69
CAM, 9, 102, 104
Cell, 4–6, 13, 61, 67, 80, 102, 104, 109, 110, 129
   delay, 63–65
   see Logic cell
   RAM, 79
Chain
   see Carry chain
   see Delay chain
Chip, ix, 1, 3, 4, 62, 74, 79, 83, 92, 131
   see ASIC
Clock
   cycle, 27, 29, 30, 32–38, 74–83, 90, 92–95
   frequency, 61, 64, 75, 77, 92
   period, 22, 65, 75, 78, 82
Coincidence, 11, 12, 97, 111, 136
Combinational logic, 4, 5, 19, 60
Common
   burst, 69, 70
   clock, 62, 69
   mode noise, 73
   offset, 110, 116
   start, 69, 70
   stop, 69, 70
   timing, 69, 70
Comparator
   magnitude, 102
   equality, 102
Compiler, 16–17
   optimization, 16–17
Compression, 53, 55–57

Computing
    power, 119, 125, 128
Consumption
    power, x, 8, 79, 80, 110
    resources, 1, 2, 12
Conversion, 13, 15, 23, 71, 74, 92, 135, 136
Converter
    see ADC
    see TDC
    LVDS-to-TTL, 86
    parallel-to-serial, 88
    serial-to-parallel, 132
Coordinate, 61, 91, 100, 102, 106, 108, 109,
        112, 119, 120, 122
Counter
    PC, 122-123
Cutoff
    frequency, 43, 44, 54–55

D

DAC, 23, 74–77, 136
    pulse width, 74, 75, 77
    pulse density, 75–77
Data, 13
    DAQ, 11
port, 7, 9, 24, 27
    storage, 13, 31
    switch, 86, 87
    transmission, 1, 88, 125
Digital clock management, 1, 131, 135, 136
Decimation, 44, 54-57
Delay, 17, 18, 27, 30, 61–65, 69, 70, 80, 83, 85,
        88, 93, 94, 120, 136
    average, 65
    see Carry chain
    cell, 63–65
    chain, 62–64, 67
    line, 53, 64, 65, 69, 93–95
    propagation, 6, 18, 19, 59–61, 63, 64, 69,
        70
    tap, 62, 65, 66
    variation, 64, 70, 94
Detector, xiii, 11, 80, 82, 86, 87, 97–99,
        107–110, 116, 126, 131–133
    hit, 80, 99, 107, 108, 110
    plane, 11, 12, 99, 100, 102, 106, 108, 109,
        112,
    SSD, 131–133
    subdetector, 86
    temperature, 37
Difference, 50, 55, 56, 64, 69, 94, 95, 116, 119,
        120, 123

phase, 59, 64
    time, 69, 135
Differential
    ADC, 74
DNL, 60, 62, 63, 66, 67
    input, 71–73
    LVDS, 131
    signal, 70, 72, 85, 86, 131
Digitizer, 77, 123
    see ADC
    see TDC
DMA, 136, 137,
Dithering, 43–47
Duty cycle, 23, 24, 37, 74, 82, 85

E

Embedded, 40
EPTSM, 131, 132
    FPGA, 131
    system, 9, 131
    microprocessor, 122
Enable
    converter, 86
    count, 22, 26, 27, 29, 30, 36, 38, 109, 110
    output, 38
    read, 79
    write, 27, 31, 32, 39, 79
Equality, 104
    comparator, 102
    double,102
    multi, 103
    single, 102
Error, 47, 66, 81, 82, 100
    binning, 70
    code, 82
    correction, 129, 130
        SEU, 17
    counter roll-over, 36
    see cross section, 128
    detection, 85, 129, 130
    EDAC, 129
    distribution, 47
    hit matching, 102
    measurement, 47, 64, 65, 66, 68, 85, 100,
        134
    see rate, 128
    RMS, 65, 66, 85
    standard deviation, 47
    track fits,12, 111–113
    truncation, 47
Event, 12, 31–33, 66, 67, 80, 86–88, 97–100,
        102, 106, 108, 109, 116, 117

building, 86–88
SEE, 125
SEFI, 126
SETI, 126
SEU, 17, 126–130
solar, 127
Experiment, 1, 59, 77, 85, 86, 97, 106, 110
   non-triggered, 77
   self-triggered, 136
   triggered, 82, 136

F

FFT, 13
FIFO, 77–82, 88, 131, 136, 137
   buffer, 78–81
   vs. pipeline, 78–82
   error, 81, 82
Filter, 43, 46
   analog low-pass, 23, 43, 74–77
      anti-aliasing, 43, 44
   digital, 46, 52–55
      anti-aliasing, 54
      CIC, 52, 53,
      see Decimation
      low-pass, 50
   notch, 75
Firmware
   TDC, 70
   pulse, 60
Flip-flop, 5, 6, 13, 18, 19, 61, 79
   D-type, 19, 32
   pipeline buffer, 79
   power consumption, 79, 80
   vs. RAM, 79, 80
   register, 32, 35, 63, 79
   SEU, 129
FPGA 1,
   applications, ix, x, 1, 16, 21–41, 59–95,
      97–113, 125, 131-136
   vs. ASIC, 3–5, 13, 18, 19, 32, 79, 92
   compiler, 16, 18
   computing, 1, 3, 13, 46–49, 51, 70, 88, 110,
      115–123
   configuration, 4, 8, 16, 21, 31, 32, 64, 66,
      69, 71, 72, 85, 86, 89, 97, 102, 110,
      125,128, 130, 135
   cost of, 1–3
   design, ix, 2, 3, 6, 11, 12, 14 -16, 18, 19, 21,
      24, 44, 59, 60–63, 73–75, 77–80, 82,
      83, 86–88, 97, 110, 125–130
   functionality, 1, 11, 13, 31, 59, 70, 78, 97,
      102, 135

initialization, 17, 22, 27, 50, 86, 122
   platform, 1, 17
   program, 2, 128
   resources, x, 1, 3–8, 11, 13, 14, 18, 19, 51,
      63, 64, 66, 69–72, 74, 85, 89, 110,
      112, 116, 125
   routing, 18, 64
   signal processing, 1, 19, 43, 54–56
   tools, 16
   transistor, 4, 13
   web site, x
Firmware, 1, 3, 21, 53
   DNL, 63
   LED, 21
   TDC, 70
Frequency
   aliasing, 43
   clock, 61, 64, 75, 77, 92
   cut-off, 43, 44, 54, 55
   fundamental, 75
   noise, 56, 76,
   operating, 6, 16, 18, 19, 33, 56
   range, 21
   RAW hazard, 33
   response, 50, 52, 54
      filter, 50, 52, 54, 75
   sampling, 43, 54
   strobe, 134
   toggling, 18, 21
Front-end, 11, 31, 131, 132
   cabling, 85, 86
   differential pair, 85, 86
   digitizer, 77, 82
   electronics, xxiii, 11, 59, 131
   error source, 81
FPGA, 31, 59, 80
   non-triggered, 77
   triggered, 80, 82

G

Galactic cosmic rays, 126, 127
Global
   command, 51
   reset, 35, 106
      RAM, 106
   trigger, 80, 98, 99

H

Harmonics, 77
Hash sorter, 9, 88, 105, 106
Hazard 32

RAW, 32, 33, 35
Histogram, 31, 33, 37, 110
    bin, 36, 110
    booking, 31–37, 67, 68
        circuit, 31, 32, 34
    DNL, 66–68
Hough transform, 108, 109
    LUT, 66
    multi-dimensional, 108, 109, 110
    online, 37
    pipelined, 33, 35
    RAM, 31, 36, 106
    reset, 35, 36, 106

I

ID
    block, 40
    status block, 41
Inequality, 100
Initialization, 17, 22, 27, 50, 86, 122
Input / output   , 1, 13–15, 39
    see Read
    see Write
Input
    ADC, 43, 44–46, 50, 71, 115
    adder, 17, 34–36
    address, 26, 27
    AND, 5
    bandwidth, 43
    buffer, 59–61, 71, 72
    clear, 27
    clock, 21, 22, 33, 85, 92
    counter, 26, 38, 109
    DAC, 74, 75, 76
    data, 7, 9, 11, 14, 16, 17, 19, 27, 47–51,
          88–90, 92, 94, 102, 115–117
    delay, 61, 63, 94
    DPF, 90
    filter, 50
    FF, 5
    FSM, 122,
I/O driver, 1
    LE, 19
    logic level, 59, 68, 71
    LUT, 4, 5, 8, 19, 60, 94, 119, 121, 122
    multiplexer, 15
    NAND/NOR, 4
    pin, 61, 71
    pipeline, 18, 33
    port, 7, 15, 16, 19, 23, 24, 26, 27
    preset, 22
    RAM, 13, 19, 31, 33, 39

RAW, 32, 33
ringing, 60
ROM, 118
sampling, 59, 73
sequence, 32, 49, 50, 117,
signal, 5, 17, 22, 28, 29, 59, 69, 72, 74
    carry chain, 62
    CC, 83
    charge, 59, 133, 134
    differential, 73
    edge, 59
    mean, 115
    pedestal, 115,
    RMS, 115
    timing, 62, 64, 68, 69, 90, 94, 97
    sliding sum, 51
TDC, 59, 66, 68, 72, 74
TMR, 17
transition, 65, 67, 94
TSO, 87
waveform, 71
Interconnect, 2, 5, 15, 15, 18, 64,
Interval, 24, 33, 38, 43, 59, 69, 82, 117, 118

K

Key number, 105, 106
    turnkey applications, 63

L

Latch, 61, 135
    SEL, 126
Latency, 26, 27, 36, 81
Light-emitting diode 21–25, 39, 75
    brightness, 23–25, 75
    clock indicator, 39
    rhythm, 21
Linearity
    DNL, 60, 62, 63, 66–68
Logarithmic
    LUT, 13
    shifter, 93, 94, 110, 117, 120
Logic, 14, 16–18, 21, 22, 59, 61, 63, 64, 69, 79,
          110, 112, 118, 123, 129
    ALU, 11, 122
    array, 18, 60–62, 67,
    block, 1, 33, 128
    carry, 63
    coincidence, 110
    cell, 5, 7, 18, 92–94
    circuit, 7, 16, 17, 18, 129
    combinational, 4, 5, 8, 19, 60

coincidence, 110, 111
element, 2, 3, 5–9, 11–15, 18, 19, 39, 56,
    59–61, 64, 67, 83, 85, 102, 105, 110,
    122, 126
function, 5–7, 13, 18, 60, 61, 63
glue, 1
level, 23, 59, 97, 131, 133
LUT, 8
    pattern, 67, 68
    sequential, 82, 83
    transition, 59, 63–68, 71, 89
    TTL, 86
Look-up table, 4–8, 13, 19, 31, 39, 60, 62–64,
    66, 67, 73, 83, 94, 115, 116, 118–121
Loop, 123
    feedback, 62
    ELMS, 122, 123,
    multi-layer, 27, 30, 121, 123
    nested, 12, 27, 30, 99, 100, 102, 106, 108,
        121, 123
    PLL, 59, 69, 83,
    single-layer, 25–28, 30, 121
LVDS, 1, 86, 131, 132, 135

M

Measurement, 12, 37, 47, 53, 55, 59, 62–64,
    67–70, 74, 97, 98, 100, 107, 108, 110,
    112, 116
    error, 47, 64, 65, 68, 85, 134
    precision, 44, 46, 47
    resolution, 63
Memory, 2, 26, 31, 32, 53, 78, 82, 120, 122
    BRAM, 128
    bin, 106
    bit, 40
    block, 5, 26, 27, 39, 64, 67, 118, 119, 128
    buffer, 51
    cell, 102, 126
    configuration, 129
    see CAM
    DMA, 136, 137
    dual-port, 88
    internal, 66
    latency, 26, 27
    LUT, 120
    output, 31
    see RAM
    SEU, 126, 130
    SRAM, 128
    updating, 28, 29
Microprocessor, 3, 5, 7, 11, 13, 25, 32, 33, 35,
    37, 39, 43, 122, 123

MICROMEGA, 106, 108
Multibit
    command, 81
    counter, 21
Multichannel deserialization, 92–94
Multidrop cable, 70
Multiequality, 103
Multi-I/O, 13
Multilayer loop, 27, 123
Multi-phase clock, 59
Multiplexer, 8, 14–16, 41
Multiplier, 5, 7, 12–14, 19, 24, 37, 38, 47–50,
    110, 112, 116, 117, 119–121
Multisampler, 60, 61, 89
Multiwire chamber, 106, 107, 110

N

NAND, 4, 83
Noise
    data transmission, 1, 131
    discrimination, 133, 135
    dithering, 44–46
    elimination, 43
    frequency, 50, 75, 76
    generation, 46
    Huffman coding, 55–57
    level, 46
    peak, 52
    pedestal, 115
    random, 117
    separation, 43
    and signal, 50, 116, 117
    source, 48, 56, 73, 74
    white, 45. 46
Non-Linearity
    DNL, 62, 63, 66, 67
NOR, 4

O

Operation, 8, 9, 11–14, 16, 21, 22, 24, 27, 31,
    32, 34, 36, 37
    arithmetic, 37
    multiplication, 7
    operating frequency, 6, 8, 16, 18, 19, 33
    sequential, 5
    synchronous, 7
Output
    accumulator, 25, 48
    ADC, 44, 45
    adder, 19, 35,
    bit, 48, 92, 94, 95

buffer, 14
CO, 24
combinational, 5
comparator, 23, 74, 131, 133, 135
counter, 17
DAC, 74–76
data, 7, 26, 47, 54, 78, 80, 81
decimation, 54
driver, 1
duty cycle, 23
edge, 59
FIFO, 79
file, 8, 61
FPGA, 54, 72, 73, 78, 88, 115
histogram, 31
LED, 21, 22
LUT, 66, 67, 118, 120, 121
memory, 26, 31
pin, 23, 71, 74
pipeline, 81
port, 7, 19, 23, 26, 27, 36, 88
pulse, 23
RAM, 33, 34, 35, 36
RC, 36,
register, 18, 38
sequence, 50
serial, 37, 38, 39

P

Personal Computer, 132
　　PowerPC, 7
Phase, 136
　　clock, 59, 61
　　data, 88, 89, 90
　　detect, 93
　　difference, 59, 64
　　DPF, 88, 89, 90, 91, 92, 94
　　drift, 90, 94
　　multi-phase clock, 59
PLL, 59, 69, 83
Pile-up 22, 53
Pin
　　FPGA, 23, 74, 86
　　DQ, 38, 39
　　ground, 21, 38
　　I/O, 14, 23, 61, 71, 73, 74
　　1-Wire bus, 39
Pipeline, 5, 6, 15, 18, 19, 33–35, 51, 59, 78–83,
　　　　86, 90, 119, 120, 122
Pixel, 87, 92
PMT, 97, 132
Port

control, 6
data, 7, 9, 24, 27
dual, 7, 9, 13, 31, 32, 39, 88
I/O, 13, 19, 26, 31
input, 15, 16, 27
output, 23, 24, 27, 36, 88
read, 31, 40
serial, 7
write, 31, 32, 36
Power
　　cable, 86
　　consumption, X, 7, 8, 13, 21, 79, 80, 110
　　cycling, 126
　　PC, 7
　　pin, 21, 39, 74, 86
　　supply, 41, 61–64, 67, 86
　　up, 21, 26, 27, 67, 86
　　usage, 2, 131
Preset, 22
Process, 14, 17, 18, 19, 33, 51
　　averaging, 46, 55
　　channel initiation, 82
　　decimation, 44, 54, 57
　　diagnostic, 41
　　error handling, 82
　　event building, 86
　　filtering, 46, 54
　　hit matching, 100, 102, 109
　　merging, 87
　　optimization, 17
　　phase following, 91
　　sequencing, 11–14
　　shifting, 94
　　synthesis, 16
　　TDC calibration, 66–68
　　tracking, 49
　　track fitting, 110
　　triplet finding, 107, 108, 110
　　zero-suppression, 77, 80, 81
Processing, 11, 17
　　ADC, 70
　　data, 43
　　digital, 46, 64
　　function, 3, 50, 88
　　pipeline, 5, 19, 35
　　signal, x, 1, 43, 50
　　time, 100, 102, 106, 108
Processing unit, 11
　　CPU, 87, 125, 126, 136, 137
Processor
　　preprocessor, 87
　　postprocessor, 86
Program, 2, 4, 21, 24, 122, 123, 129

control functions, 122
counter, 122, 123
firmware, 21
in FSM, 122
loops, 122, 123
re-program, 24
Pulse
density, 75, 76
display, 22
edge, 60
filter, 60, 61
height, 1
input, 59
internal, 24
length, 22, 24
output, 23
ST, 26, 27, 30
time, 37, 39, 118
train, 22, 38, 43, 67
width, 22, 60, 74, 75

R

RAM, 1, 4–9, 13, 15, 19, 32–36, 40, 51, 79, 88, 94, 105, 106, 129, 131, 136
BRAM, 128
cell, 79
distributed, 5, 8, 9
dual port, 7, 13, 31, 39
ethernet, 137
frequency, 19
histogram, 31–34, 36
index, 36
port, 6, 36, 40
reset, 35, 106
single-loop sequencer, 28
SRAM, 128
synchronous, 87
trigger, 80
Ramp
ADC, 7–73
Read, 32, 37, 40, 102, 122, 131, 136
RA, 27, 31, 32, 35, 36, 79, 81
after write, 32
command, 39
RE, 79,
port, 31, 40
sequence, 39, 40
simultaneous with write, 31, 35
strobe, 131, 134–136
Read-only memory, 7, 39, 40, 118
in microsequencer, 122, 123
and LUT, 7, 118

Readout, 7
ASIC, 132
data, 35
pixel, 92
PTSM, 132
RAM, 40
path, 41
serial number, 38–41
simultaneous, 136
Real
events, 66, 67, 86
hit, 77, 78
time, 64
Redundancy
C5 scheme, 85
CRC, 38
measurement, 12, 112
SEU, 129
TMR, 129
Reset, 38, 123, 135
cycle, 36
C5 decoder, 83, 85
global, 35, 51, 106
histogram, 35, 36, 37, 106
microsequencer, 123
system, 66
Response, 45, 74
frequency, 50, 52, 54
time, 65, 74
Routing, 18, 64

S

Single-event effect, 125, 126
SEFI, 126
SEL, 126
SET, 126
SEU, 17, 126, 128, 130
Sequence, 24–27, 30, 32, 39, 43, 45, 47, 49–51, 55, 66, 115, 117, 121, 122, 123
address, 26
control, 24, 26, 66, 122
ELMS, 122, 123
memory updating, 29
microsequence, 22, 24, 25, 122, 123
multilayer, 27
read, 39, 40
sequencer, 26
single-loop, 28
two-layer, 29
write, 40
Session, 36
booking, 35, 36

Shape
    pulse, 53, 133, 134
    ramp, 72, 73
    response, 52
    saw-tooth, 102
Shaper, 43, 44, 56, 133
Signal
    ADC, 43, 45
    analog, 1, 71
    band-limiting, 43
    bandwidth, 43, 54
    CEA, 27
    clock, 21, 29, 62, 66, 69, 82, 135
    count
        control, 27
        enable, 22, 24, 27, 29, 30, 109, 110
    counter enable, 26
    CO, 24
    combinatorial, 5
    component, 43
    configuration, 86
    continuous, 43
    differential, 70, 72, 131
    digital, 1, 43
    DQ, 39
    DSP, 123
    DV, 31, 38, 56, 59, 80, 85
    frequency, 50
    input, 5, 17
    length, 133, 134
    logic function, 5
    LVDS, 131, 132
    mixed, 1, 131
    noise, 45, 50
    parallel, 133
    path, 60
    periodic, 75
    PMT, 97
    propagation, 19
    processing, X, 1, 43, 50, 123
    push, 79, 80
    RE, 79
    reference, 69, 70
    reflection, 60
    reset, 35, 36, 106
    ringing, 60
    sequence, 115
    scintillator, 135
    shape, 53
    slow, 72, 74
    ST, 27, 30
    start/stop, 69
    stretched, 97

    TDC, 59, 66, 69
    timing, 62, 64, 68, 69, 70, 82, 97
    TOT, 135
    trigger, 131
    voice, 77
    WE, 27, 31, 32, 39, 7
Silicon
    area, 1, 110
    EPTSM, 131, 132
    resources, ix, x, xiii, 2, 11, 12, 13, 14, 15,
        19, 51, 56, 79, 92, 100, 102, 105, 110,
        112, 115–118, 122
    serial number, 40
    SOI, 128
    strip sensor, 131–133, 136
    usage, 100
Slope
    dual slope ADC, 72
    ramping slope ADC, 72
    single slope ADC, 71–74
    track, 106, 111, 112
Software
    averages, 46
    compiler, 21
    design, 128
    FPGA design, 14, 18, 61
    mitigation, 125, 127
    pairing, 100
    TMR, 129
    track fitting, 110
    trigger, 97
    triplet finding, 12, 108
Sorter
    Hash sorter, 9, 88, 105, 106
    regular, 106
Stage
    analysis, 97, 110
    analog, 46
    cascaded, 52
    configuration, 5, 8
    delay-line, 94, 95
    design, 19, 44,
    implementation, IX
    initialization, 86
    logarithmic shifter, 94
    operating, 8
    pipeline, 5, 6, 18, 19, 33, 34, 59, 120
    planning, 7, 19
    preprocessing, 87
    processing, 5
    sampling, 61
    synthesis, 17
    trigger, 99, 110

triplet-finding, 108
zero-suppression, 77, 81
Start, 83,134
common, 69, 70
ST, 27
starting point, 76, 122
starting time, 72, 81, 136
start-up, 51
Stop
band, 50, 57
common, 69, 70, 134
time, 136
Synchronization, 69, 82
Synthesis, 16, 17

T

TDC, 17, 59
ASIC, 62, 69
bin number, 66, 68
bin width, 62, 63, 67, 68
calibration, 63, 64, 65, 66
delay-line, 65
bin-by-bin, 65, 66
error, 65, 66
LUT, 66, 67
carry chain, 17, 64
C5, 82
common timing reference distribution, 69, 70
burst, 69, 70
jitter, 70, 85
mean timing, 70
start/stop, 69
delay-chain, 62, 63, 64, 67
DNL, 62–68
DPF, 89
multi-phase clock, 59
multi-sampling, 59, 60, 61
propagation delay, 64
sensitivity, 61–64
resolution, 63, 68
TDC block inside ADC, 71–74
timing off-set, 69
trigger primitives, 97, 98
wave union, 67–69
zero-suppression, 77–81
TS, 77, 78, 80
Temperature
calculation, 37
detector, 37
digitization, 37, 38
measurement, 37

sensitivity, 61–64, 67, 70, 88, 90
serial data, 41
Time
average, 70
of arrival, 59, 62, 64, 68–70, 80, 97, 116–118, 131
bit, 89, 90, 92, 94
calibration, 65, 72
of the center, 66, 67
coarse, 59, 60, 77, 81
coincidence, 136
of comparator high/low, 73–75
of compilation, 2
of computation, 11, 100, 102, 106, 116, 119, 120
of FFT, 13
constant, 21, 49, 50, 74
correlation, 1
counter, 60, 77
critical, 18
delay, 65
difference, 69, 135
domain, 97
encoder, 62
estimation, 117, 118
of execution, 2, 12, 100, 108
of TTF, 12
fine, 59, 62, 81
of hit, 106, 108
interval, 24, 43, 69
TMP03/04, 37, 38, 39
mean, 70, 71, 80
measurement, 59
period
LED, 23
DAC, 75
in digitizers, 77
precision, 116
pulse high/low
TMP03/04, 37
DS2401, 39
C5, 82
DPF, 90
pulse offset, 118
real, 64
recover, 39
reference, 70
response, 65
resolution, 98
sampling, 117
slots, 39, 81
stamp, 77, 78, 80–82, 87, 136, 137
starting, 72, 81, 136

stop, 136
step, 121
of TDC, 67
tick, 24
transition, 71, 72, 89
window, 108, 129
Time-of-Flight
    detector, 97, 98
    counter, 97
    hit position, 98
    resolution, 97
    trigger primitives, 98
Time-over-Threshold
    binary, 131
    counters, 1
    EPTSM, 131, 133
    FPGA functions, 136
    error, 134
    vs. input charge, 133–135
    PMFE ASIC, 133
    Read cycles, 135,
    saturation, 133
Timing
    burst, 70
    calibration, 66
    common, 69, 70
    conditions, 18
    constraints, 18
    critical signal path, 60
    critical logic element, 60
    critical buffer and FF, 60
    diagram
        ending sequence, 30
        memory updating sequence, 29
        serial number readout, 40
        single-loop sequencer, 26
        two-layer nested loop, 30
    distribution, 70
    jitter, 70
    mean timing scheme, 70, 71
    measurement, 53
    reference, 69, 70
    relative, 17, 62, 69, 131
    requirements, 18
    resolution, 59, 61, 70
    signal transmission, 82
    tick, 24
    window, 81
Time projection chamber
    triplet finding, 106, 108, 110
Total ionizing dose, 125, 127
Track
    circular, 106, 107, 110

configurations, 108
constraint, 110
coordinate, 99
curved, 12, 111, 112
detection, 99, 131, 132,
finding and Hough transform, 108–110
fitting, 12, 110–113
from decay, 99
hits, 99, 102, 112
hit time, 80, 106
in TPC, 106–108
length, 113
parameters, 111
projection, 12
recognition, 106
reconstruction, 99
roads, 109
segment, 12, 87, 106, 108
segment finder, 12, 110
    Hough transform, 108
separation, 97
straight, 106–108, 110
Tracking
    averages, 50
    voltages, 74
    phase, 89
    progress, 49
    smoothness, 50
Transceiver
    built-in, 88
    and DPF, 88
    serial, 7, 8, 88
TTL, 86
Transition
    arrival time, 69
    bin number, 66, 68
    common timing, 69
    during ramp, 71
    detection, 59, 60, 89, 90, 92
    regulation, 59, 60
    false, 60
    input logic, 59, 63, 65, 66,
    location, 90
    logic transition, 67, 68
    message, 83
    phase, 94
    position, 17, 62, 65, 68, 90
    separation, 68
    time, 71, 72, 89
    wave union, 67, 68
    valid, 60
Trigger
    acceptance command, 82

and DAQ, 11, 78, 80, 81, 86, 87
  command, 81
  decision, 135, 136
  during data taking, 86
  event, 80
  external, 131, 135
  front-end, 80
  global, 80, 98, 99
  high-level, 110
  latency, 81
  level 1, 80, 81
  non-trigger, 77
  number, 80
  online, 100, 108, 117
  packet data header, 80
  pixel system, 97
  primitives, 80, 81, 97, 98
  self-trigger, 136
  software trigger, 97
  system, 81, 86, 97, 99
Tri-state
  buffer, 14, 15
  bus, 14, 15
  pin, 86
Triplet, 107
  finding, 106 -108, 110
    Hough transform, 108, 109
TTF, 12, 110, 111

U

Upset
  SEU, 17, 126, 128, 130

V

VHDL, x, 2, 10

Vertex, 87

W

Write
  address, 27, 31, 32, 40, 79, 102
  bit, 39
  byte, 39
  command, 39
  DQ, 40
  enable, 31, 32, 39, 79, 80
  histogram reset, 35
  operation, 31
  port, 31, 32, 36, 40
  sequence, 40
  simultaneous with read, 31

X

XILINX, 5, 7, 8, 128, 131, 132

Y

Y2K, 77

Z

Zero
  net current, 131
  low-pass filter, 54
  overhead looping, 123
  suppression, 59, 77, 80–82, 136
  ZBT, 87, 88

Printed and bound by CPI Group (UK) Ltd, Croydon, CR0 4YY

18/10/2024

01776236-0002